The Large Hadron Collider: a Marvel of Technology

edited by Lyndon Evans

EPFL PRESS

Distributed by CRC Press

Taylor and Francis Group, LLC
6000 Broken Sound Parkway, NW,
Suite 300,
Boca Raton, FL 33487

Distribution and Customer Service
orders@crcpress.com

www.crcpress.com

Library of Congress
Cataloging-in-Publication Data
A catalog record for this book
is available from the Library
of Congress.

The publisher extends his thanks
to Mr. Anton Vos, who provided
editorial guidance in the early phases
of this project; to our Series Editor
Professor Philippe-André Martin
(EPFL); and to Dr. Mario Campanelli
of CERN, for his invaluable editorial
feedback during the production
of the book.

The graphical layout of the book
and cover design are by Paola
Ranzini Pallavicini of Studio Pagina,
Milan, Italy.

The photographs used in this book,
unless otherwise noted, are
reproduced with permission from
CERN. Special thanks to Maximilien
Brice, Claudia Marcelloni and their
colleagues at the CERN Press Office
who took many of these photographs.
Photographs by Peter Ginter,
Antonio Saba and Michael Hoch are
also reproduced with permission.

The EPFL Press is an imprint owned
by the Presses polytechniques
et universitaires romandes, a Swiss
academic publishing company whose
main purpose is to publish
the teaching and research works
of the Ecole polytechnique fédérale
de Lausanne (EPFL).

Presses polytechniques
et universitaires romandes
EPFL – Rolex Learning Center
Post office box 119
CH-1015 Lausanne, Switzerland
e-mail: ppur@epfl.ch
Phone: +41 (0) 21 693 21 30
Fax: +41 (0) 21 693 21 30

www.epflpress.org

Printed in Italy

© 2009, 2014, first edition, CERN and EPFL Press
ISBN 978-2-940222-34-6
ISBN 978-1-4398-0401-8

The publisher, editor and authors thank the industrial sponsors of this book for their generous support:

ALST(O)M

AFT microwave GmbH
www.aft-microwave.com

Bayards Aluminium Constructies BV
www.bayards.nl

buntmetall amstetten GmbH
www.buntmetall.at

CES Creative Electronic Systems SA
www.ces.ch

Cogent Power Ltd.
www.cogent-power.com

REEL S.A.S.
www.reel.fr

RI Research Instruments GmbH
www.research-instruments.de

A. SILVA MATOS, SA
www.asilvamatos.pt

ZEG S.A.
www.zeg.pl

Qualitech Ltd. is an independent company, specialised in non-destructive testing (accreditedby SAS in conformance with ISO/IEC 17025) and destructive testing. Stationary tests are performed in the bunkers for X-ray tests at Maegenwil, Altenrhein and Biel. Mobile operations are made outside. Radiography is made using X-ray tubes and isotopes. Other methods, e.g. ultrasonic and eddy current testing; testing of surface cracks and cables; and special tests are offered, too. Personnel certifications enable approval testing of welders in accordance with ISO/IEC 17024. Inspectors are verifying the fulfilment of quality specifications of the clients and their documents worldwide.

www.qualitech.ch

Alstom is a global leader in the world of power generation and rail infrastructure and sets the benchmark for innovative and environmentally friendly technologies. Alstom builds the fastest train and the highest capacity automated metro in the world, and provides turnkey integrated power plant solutions, equipment and associated services for a wide variety of energy sources, including hydro, nuclear, gas, coal and wind. The Group employs more than 81,000 people in 70 countries, and had orders of € 24.6 billion in 2008/09.

www.power.alstom.com

Table of Contents

1.0

THE LARGE HADRON COLLIDER: AN INTRODUCTION

Lyndon Evans

On the 10th July 1908, in his laboratory in Leiden, Heike Kamerlingh Onnes became the first person in the world to liquefy helium. Between 6.30 p.m and 7.30 p.m he succeeded in producing 60 ml of liquid, enough to fill a small teacup, beating Sir James Dewar to the race to liquefy the last remaining gas. He measured its temperature to be −269 °C, the lowest temperature ever achieved on earth.

The preferred temperature scale for scientists, particularly for cryogenic applications, is the Kelvin scale. It has the same size of graduation as the Celsius scale but its zero is shifted; zero Kelvin corresponds to −273 °C, the absolute zero of temperature below which no material can be cooled. It got its name from William Thompson, a pioneer of the science of thermodynamics, who, when he was ennobled (for work on transatlantic telegraphy), took the title Lord Kelvin after the name of the river that meanders through Kelvingrove park in front of Glasgow University. The most famous temperature scale in science is named after a Scottish river! The temperature of the liquid on this scale measured by Kamerlingh Onnes was precisely 4.22 K.

He tried to solidify the helium by reducing the temperature even further using an old trick. It is well known (at least to the English) that tea tastes better at sea level than on the top of a high mountain. This is because the lower atmospheric pressure at high altitude makes the water boil at a lower temperature. By reducing the pressure above the helium liquid using giant vacuum pumps, he was able to further cool the liquid to 1.5 K but could not freeze it. In fact, he missed a very important discovery. Surprisingly, it took a further 20 years to realize that at 2.17 K, the liquid undergoes a phase transition (just as water does at 273 K). At precisely this temperature, these days called the lambda point, the liquid becomes a macroscopic quantum state, exhibiting bizarre properties that can only be explained by the laws of quantum mechanics. Kamerlingh Onnes produced superfluid helium without knowing it. Looking back, it is quite surprising that he missed it because he could see the liquid in his glass vessel, and the transition to the superfluid state is visually quite impressive. Above the lambda point, the liquid boils violently. At the transition temperature, the boiling suddenly stops because one of the characteristics of the superfluid state is a very high thermal conductivity. The liquid is unable to support the temperature gradient that leads to bubbling; it becomes totally quiescent.

Kamerlingh Onnes soon used liquid helium to cool down other materials in order to measure their properties at very low temperature. In 1911, he

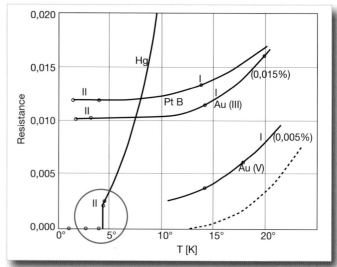

discovered that the resistance of solid mercury abruptly disappeared at 4.2 K, a property that he dubbed "superconductivity". In later years, a large number of metals were found to exhibit the same property, the ability to sustain an electrical current without loss at low temperature. Modern superconductors can carry very large currents. The most effective of these readily available is an alloy of Niobium and Titanium (NbTi).

A century of progress

Almost exactly 100 years later, these two discoveries, superconductivity and superfluidity, have been brought together as the two pillars on which the design of the largest and most complex scientific instrument ever built rests. The Large Hadron Collider (LHC) at the European Centre for Particle Physics (CERN) will allow scientists to delve even deeper into the secrets of nature. The construction of the LHC and its detectors has been a monumental effort spanning almost 15 years and involving scientists and engineers from all over the planet.

The LHC is the latest and most powerful in a series of particle accelerators that allows scientists to probe the structure of matter at its tiniest dimension. It consists, in fact, of two particle accelerators located in a 27-km circular tunnel, 100 meters underground. Two counter-rotating beams of protons (the nuclei of the hydrogen atom, from the family of particles called *hadrons*, hence its name) are accelerated to the unprecedented energy of 7 tera-electron volts (TeV) and brought into collision in four huge detectors that capture the debris from the collisions. It can also accelerate other projectiles. In particular, there is an approved program of physics with heavy (lead) ions during a modest part of the machine operating time. The beams are guided around their circular orbits by powerful superconducting magnets cooled in a bath of superfluid helium. However, compared with the few grams of liquid first obtained by Kamerlingh Onnes, the LHC requires 130 tons to cool it.

The LHC is a wonder of modern technology. The LHC detectors are no less so. In the following chapters, the main challenges encountered during their construction are described and some of the fundamental questions in science to be addressed are discussed. But first, as with any large and expensive scientific project, there were a number of political hurdles to be crossed.

1. The original refrigerator of Kamerlingh Onnes in Leiden University. With it, he managed to liquefy 60 ml of helium in an hour. One LHC plant liquefies 4000 l/h.

2. Resistivity of various materials as their temperature approaches absolute zero. The first superconductor discovered in 1911 by Kamerlingh Onnes was mercury. It loses its electrical resistance entirely at about 4.2 K.

Approval of the LHC

The LHC had a difficult birth. Although the idea of a large proton-proton collider at CERN had been around since at least 1977, the approval of the Superconducting Super Collider (SSC) in the United States in 1987 put the whole project into doubt. The SSC, with a centre-of-mass energy of 40 teraelectron volts (TeV) was almost three times more powerful than what could ever be built at CERN. It was only the resilience and conviction of Carlo Rubbia, who shared the 1984 Nobel Prize in physics for the discovery of the W and Z bosons that kept the LHC project alive. Rubbia, who became Director General of CERN in 1989, argued that, in spite of its disadvantage in energy, the LHC could be competitive with the SSC by having *luminosity* (basically the rate of production of collision events) an order of magnitude higher than could be achieved with the SSC, and at a fraction of the cost. He also argued that the LHC would be more versatile. As well as colliding protons, it would be able to accelerate heavy ions to world-beating energies at little extra cost.

The SSC was eventually cancelled in 1993 after a series of cost overruns that escalated the expected cost from the $4.4 billion approved to more than $11 billion. This made the case for the building of the LHC even stronger, but the financial climate in Europe at the time was not conducive to the approval of a large project. CERN's largest contributor, Germany, was struggling with the cost of reunification and many other countries were trying to get to grips with the problem of meeting the Maastricht criteria for the introduction of the single European currency.

During the course of 1993, an extensive review was made in order to reduce the cost as much as possible, although a detailed cost estimate was par-

3. An outline of the position of the 27-km LHC ring. The airport of Geneva can be seen in the foreground. The dotted line is the border between Switzerland and France. The LHC tunnel is approximately 100 meters underground: nothing can be seen from the surface.

ticularly difficult to make since much of the research and development on the most critical components was still to be done. In December 1993, a plan was presented to the CERN Council to build the machine over a ten-year period by reducing the other experimental program of CERN to the absolute minimum, with the exception of the full exploitation of the Large Electron Positron (LEP) collider, which was the flagship machine of the decade.

Although the plan was generally well received, it became clear that two of the largest contributors, Germany and the United Kingdom, were very unlikely to agree to the budget increase required. They also managed to get Council-voting procedures changed from a simple majority to a double majority, where much more weight was given to the large contributors so that they could keep control.

On the positive side, after the demise of the SSC, a US panel on the future of particle physics recommended that "the government should declare its intentions to join other nations in constructing the LHC". Positive signals were also being received from India, Japan and Russia.

In June 1994, the proposal to build the LHC was made once more to Council. Seventeen member states voted to approve the project. However, because of the newly adopted double voting procedure, approval was blocked by Germany and the UK, who demanded substantial additional contributions from the two host states, France and Switzerland, claiming that they obtained disproportionate returns from the CERN budget. They also requested that financial planning should proceed under the assumption of 2% annual inflation, with a budget compensation of 1%, essentially resulting in a 1% annual reduction in real terms.

In order to deal with this new constraint, we were forced to propose a "missing magnet" machine where only two thirds of the dipole magnets needed to guide the beams on their quasi-circular orbits would be installed in a first stage, allowing the machine to run with reduced energy for a number of years, eventually upgrading to full energy. This would have been a very inefficient way of building the machine, costing more in the long run but saving some 300 million Swiss francs in the first phase. This proposal was put to Council in December 1994. The deadlock concerning extra host-state contributions was broken when France and Switzerland agreed to make extra voluntary contributions in the form of a 2% annual inflation adjustment, compared with the 1% adjustment from the other member states. The project was approved for two-stage construction, to be reviewed in 1997 after the size of the contribution offered by non-member states interested in joining the LHC program would be known.

There followed an intense round of negotiations with potential contributors. The first country to declare a financial contribution was Japan, which became an observer to the CERN Council in June 1995. The Minister attended the Council meeting personally, and a very interesting ceremony took place. There is a Japanese custom that at the start of collaboration, one eye of a Japanese Daruma doll is painted. This was duly done by Minister Yosano and the then Director General of CERN, Chris Llewellyn Smith. At the successful termination of a project, the other eye is filled in.

Thirteen years later, the representative of the Japanese Government at the LHC Inauguration Ceremony, senior vice-minister Toshio Yamauchi, representing Minister Yosano completed the job by painting the other eye in the presence of the serving Director General Robert Aymar, Chris Llewellyn Smith and Swiss President Pascal Couchpain, thereby signaling the end of a long and fruitful collaboration in the construction of the LHC.

4. Minister Yosano and the former Director General of CERN, Chris Llewellyn Smith holding the Daruma doll with one eye painted. Between them is the President of the CERN Council and former French Minister of Research, Hubert Curien.

5. To mark the LHC Project completion, the second eye of the Daruma doll was painted by Mr Toshio Yamauchi (left), senior vice-minister of education in Japan in the presence of Director General Robert Aymar (right), Chris Llewellyn Smith (second right) and Swiss President Pascal Couchepin.

The declaration from Japan was quickly followed by India and Russia in March 1996 and by Canada in December.

A final sting in the tail came in June 1996 from Germany who unilaterally announced that, in order to ease the burden of reunification, it intended to reduce its CERN subscription by between 8% and 9%. Confining the cut to Germany proved impossible. The UK was the first to demand a similar reduction in its contribution in spite of a letter from the UK Minister of Science during the previous round of negotiations stating that the conditions are "reasonable, fair and sustainable." The only way out was to allow CERN to take out loans, with repayment to continue after the completion of LHC construction.

In December 1996 Council, Germany declared that "a greater degree of risk would inevitably have to accompany the LHC." The project was approved for single-stage construction with the deficit financed by loans. It was also agreed that the final cost of the project was to be reviewed at the half-way stage with a view to adjusting the completion date. With all contingency removed, it was inevitable that a financial crisis would occur at some time, and this was indeed the case when the cost estimate was revised upwards by 18% in 2001. Although this was an enviable achievement for a project of such technological complexity and with a cost estimate from 1993 before a single prototype had been made, it certainly created big waves in Council. CERN was obliged to increase the level of borrowing and extend the construction period (which was anyway necessary on technical grounds for both the machine and detectors).

In the meantime, following the recommendation of the US panel, and in preparation for a substantial contribution, The US Department of Energy, responsible for particle-physics research, carried out an independent review of the project. They found that "the accelerator-project cost estimate of 2.3 billion in 1995 Swiss francs, or about $2 billion U.S., to be adequate and reasonable". Moreover, they found that "most important of all, the committee found that the project has experienced and technically knowledgeable management in place and functioning well. The strong management team, together with the CERN history of successful projects, gives the committee confidence in the successful completion of the LHC project." In December 1997, at a ceremony in Washington in the splendid Indian Treaty Room of the White House Annex, an agreement was signed between the Secretary of Energy and the president of the CERN Council. More than 1300 American physicists are users of CERN today, a remarkable number for a non-member state.

After a shaky start and a mid-term hiccup, the project has proceeded reasonably smoothly to completion. The LHC is a fine example of European collaboration and leadership in science.

7

The LHC is a machine of discovery. It will take us into a new energy regime where, through the famous relationship $E = mc^2$, new particles of greater mass, inaccessible to existing machines will be produced. It is not known what the LHC will discover, but there are a number of fundamental questions in physics on which the machine will certainly shed light. The four detectors of the LHC are each designed to answer these questions.

The origin of mass

Mass or weight (mass under the influence of a gravitational field) is such a familiar concept that we take its existence for granted. The structure of matter at the most fundamental level has been elucidated over the last 35 years by experiments at CERN and at other laboratories around the world. These experiments have revealed an astonishingly simple picture. All matter is made up of a small number of elementary particles (six quarks and six leptons) held together by forces mediated by a small number of "force" particles. The most familiar of these is the photon, the particle of light. The photon has no mass but this does not mean that it is useless. It brings the energy from the sun that allows life on earth, it allows us to see, and where would we be today without TV or mobile phones? On the other end of the spectrum, the particles that mediate the weak nuclear force (the W and Z bosons discovered at CERN in 1983), which are responsible for the burning of the sun, are very heavy, weighing respectively 80.4 GeV and 91.2 GeV, a little less than an atom of silver. The fundamental mechanism of how particles acquire mass and why there is such a large difference between them is not understood. The most promising theory predicts the existence of a particle called the Higgs boson (there may be more than one) which is responsible for the process that gives mass to the other particles. If the Higgs exists, it is very heavy. It must be heavier than 113 GeV or it would have been seen already at the CERN large electron-positron collider (LEP), the previous CERN flagship accelerator. On the other hand, fundamental arguments require that it is less than about 850 GeV . The LHC is designed to cover the whole energy range; if the Higgs exists, the LHC will find it. The two largest detectors, ATLAS (an acronym for A Toroidal LHC ApparatuS) and CMS (the Compact Muon Solenoid detector) are general-purpose detectors capable of observing the unexpected, but they are especially sensitive to all possible manifestations of the Higgs boson. These experiments are discussed in detail in Chapters 5.1-5.3.

Matter-antimatter asymmetry

All matter particles have antimatter cousins which can be created in our accelerators. These are particles of the same mass but opposite electric charge. When matter and antimatter meet, they annihilate one another, converting into radiation. If there were perfect symmetry between matter and antimatter, then during the early big bang, these annihilations would have taken place leaving a universe with only photons, no place for us! The LHCb detector ("b" for the Beauty Experiment) is designed to study this very subtle asymmetry; the experiment is described in Chapter 5.5.

Dark matter and dark energy

The first person to postulate the existence of a vast unseen form of matter was Swiss astrophysicist Fritz Zwicky in 1933. In the late 1960's and early 1970's, solid experimental evidence began to emerge from the work of Vera Rubin, a young American astronomer. Rubin and colleagues measured the rotational velocities of galaxies as a function of distance from the galactic centre using spectroscopic techniques. They found that, instead of dropping off with increasing distance, most of the stars are orbiting at roughly the same speed. The only way that this can be explained is that the density of matter was constant far beyond the visible galaxy. Another equally mysterious effect was discovered in 1998, when the LHC was well into construction. Measurements of the recession speeds of distant supernova have produced evidence that the expansion of the universe is accelerating. To explain this, a new kind of energy, "dark energy" has been postulated. It is now thought that this invisible dark matter and dark energy make up for about 96% of the total mass; only 4% of the universe is observable.

These two phenomena are examples of the convergence of particle physics and cosmology. The LHC will provide a laboratory environment where it may be possible to elucidate their cause.

The quark-gluon plasma

By colliding beams of lead ions, the LHC will be able to produce a state of matter that only existed a few millionths of a second after the big bang. The properties of the so-called quark-gluon plasma can be studied in detail in the specially built ALICE detector (an acronym for A Large Ion Collider Experiment), presented in Chapter 5.4.

A brief history of CERN colliders

From its foundation in the 1950s until the late 1960s, particle physics research at CERN was done in a way similar to that Rutherford used at the beginning of the 20th century when he discovered the atomic nucleus by bombarding a thin foil target with energetic alpha particles (the nucleus of the helium atom) from radioactive decay. In CERN's early accelerators, beams of protons (the hydrogen nucleus) replaced the alpha particles as projectile. They could be accelerated to much higher energy and could be made to collide with the nucleons in any selected target material. Now, when a high-energy proton collides with a stationary proton or neutron in a target, new particles can be created by the conversion of energy into mass according to the famous Einstein relationship. However not all of the energy of the incoming projectile is available due to the conservation laws of energy and momentum. As a consequence, the available energy for new particle production only increases very slowly, as the square root of the energy of the incoming proton. For example, in the 450 giga-electron volt (GeV) Super Proton Synchrotron (SPS) at CERN operating in this "fixed target" mode, only about 30 GeV is available for making new particles. On the other hand, if the two particles can be made to collide head-on, each with 450 GeV, the full 900 GeV is available. This is equivalent to the real life observation that the damage is much worse if two cars collide head-on with a given velocity than if a car struck a stationary vehicle with the same velocity. In the second case, much of the energy is dissipated in pushing the stationary car forward.

These colliding beam machines (storage rings), with two beams of particles circulating in opposite directions and colliding at a point on the circumference where particle detectors could be placed were the dream of accelerator builders in the late 1950s. In the early 1960s the first machines started to appear at Stanford in the US, Frascati in Italy and Novosibirsk in Russia. Instead of protons, these machines collided leptons (electrons or positrons). One great advantage in using leptons is that, when bent on a circular orbit, they emit light (synchrotron radiation). The dynamics is such that the emission of this radiation has a natural damping effect on the transverse dimensions, concentrating the particles into a very intense beam, essential if there is to be a reasonable probability of two particles colliding instead of the beams just passing through each other like two clouds. It is also desirable that the beams can circulate for many hours while data can be collected. During this time the particles are subjected to perturbations due to imperfections in the guide field or the electromagnetic field of the other beam that can drive them unstable. Synchrotron radiation also plays an important role in combating these external perturbations due to its natural damping effect. However, the emission of synchrotron radiation makes the particles lose energy, which has to be replaced by the acceleration system. Essentially, the beams have to be permanently accelerated in order to keep them at constant energy. The energy lost each revolution increases dramatically (with the fourth power) as the energy of the machine increases, eventually making it impossible for the accelerating system to replace it. In spite of its usefulness, synchrotron radiation naturally limits the maximum achievable energy of the machine. The way around this is to revert to particles that emit much less radiation.

Proton storage rings
In the late 1960s, a very bold step was taken at CERN with the construction of the first proton storage rings, called the Intersecting Storage Rings (ISR), which started operation in 1969. The advantage of protons is that they do not emit synchrotron radiation of any consequence since the energy loss per revolution varies as the inverse fourth power of the mass of the particle, and protons are 2000 times heavier than electrons. The disadvantage is that they have to operate without the benefit of the strong damping provided by synchrotron radiation. Indeed, many accelerator physicists doubted that proton storage rings would work at all.

In the end, the ISR was a big success and an essential step on the road to the LHC. The machine eventually reached 31 GeV per beam, compared with the few GeV available from the lepton beams at that time. The accelerator physicists learned how to build proton storage rings that overcame the lack of synchrotron radiation damping. The experimentalists learned how to build detectors that worked in the difficult environment of a proton-proton collider.

6. The Intersecting Storage Rings was the first proton-proton collider. The two separate rings can clearly be seen. In the ISR, the accelerator physicists learned how to store proton beams. The experimentalists learned how to build detectors.

Another disadvantage of using hadrons (protons and antiprotons) is that, unlike leptons, they are composite objects. Each proton contains three more fundamental particles (quarks) held together by gluons. Each quark carries, on average, about one fifth of the hadrons' energy. The rest is stored in the other quarks and the gluon field. When two quarks collide, the exact collision energy is not known a priori. It must be measured in the detectors by *calorimetry*, a technique that measures the energies of all the created particles. In addition, there is a very large background of unwanted events due to "soft" collisions of the gluon fields. In fact, many physicists were initially skeptical about our ability to dig out rare events from this large background.

Construction of the LEP

For these reasons, it was decided that the next machine for CERN would be LEP, the Large Electron Positron collider. In order to minimize the effect of synchrotron radiation it was necessary to build a very large, 27-km circumference ring. Even so, the maximum energy of LEP was limited to around 100 GeV, at which point it was radiating away a substantial fraction of its energy each revolution. Although LEP produced an enormous amount of precision data, it came to the end of its useful life when it hit the synchrotron radiation barrier. It was shut down in 2001 to make way for the LHC. The way to higher energies was once more to revert to protons as projectiles. The LEP tunnel is the major piece of real estate inherited by the LHC.

The final step on the road to the LHC was taken during the long period of LEP construction. During this time, Carlo Rubbia proposed that the Super Proton Synchrotron (SPS), built in the 1970's as a "fixed target" machine, could be turned into a hadron collider using the newly discovered technique of accumulating and cooling antiprotons produced in CERN's oldest machine, the CERN Proton Synchrotron (PS). Since protons and antiprotons have the same mass but opposite charge, they could be accelerated in opposite directions in the single vacuum chamber of the SPS. Collisions at 273 GeV per beam produced the first W and Z bosons, the mediators of the weak nuclear force responsible for radioactive decay. The Nobel Prize was awarded to Rubbia and van der Meer (who developed the cooling method of antiprotons) in 1984.

The Proton-Antiproton collider (PPBAR) also provided the essential remaining information needed for the design of the LHC and its detectors.

7. The Large Electron-Positron collider. LEP ran from 1989 to 2000. The maximum energy achieved was 104.5 GeV per beam. At that energy, it was losing more than 5 GeV per turn due to synchrotron radiation. It was dismantled in 2001 to make way for the LHC.

8. In January 1983, the first unambiguous signal for the W boson was obtained in the UA1 detector at CERN (colored track). Beams of protons and antiprotons were brought into collision in the CERN Super Proton Synchrotron at 273 GeV. This picture also illustrates the large background that must be dealt with in hadron collisions.

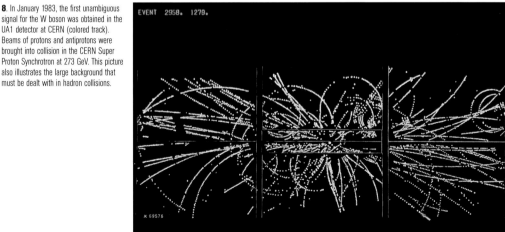

EVENT 2958. 1279.

× 69576

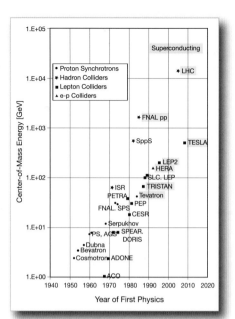

9. The history of colliders.
In squares are the lepton machines; in stars, the hadron machines, three of the four constructed at CERN. The energy available in the quarks or gluons is about one fifth of the beam energy.

For the LHC machine it elucidated the main factors that would limit the performance of the LHC, and the two detectors UA1 and UA2 served as prototypes for the much larger LHC detectors. Indeed, the nucleus of the teams designing ATLAS and CMS comes from these earlier collaborations.

The path of the protons

An essential feature of the LHC in terms of cost reduction is that the CERN infrastructure, with 50 years of investment, is used to produce the beams that eventually collide in the LHC. Without this, the cost of the project would have doubled. No single machine can accelerate the beam all the way up to 7 TeV. It needs a cascade of accelerators, all working in tandem.

Protons are the positively charged nuclei of the hydrogen atom. They are created in an ion source called a *duoplasmatron* from which they are extracted with an energy of 50 kilo-electron volts (KeV). The next step in their journey to the LHC is through a 35 meter long linear accelerator (Linac) where their energy is increased to 50 mega-electron volts (MeV). Originally, the beam was injected directly into CERN's oldest machine, the 100 meter radius Proton Synchrotron (PS), built in 1959, but in 1972 a booster synchrotron (PSB) was inserted between the Linac and PS to improve its performance. The PSB is one quarter of the circumference of the PS and contains four superposed rings to allow filling of the whole PS circumference in one pulse.

The beam is accelerated to 1.4 giga-electron volts (GeV) in the PSB and then transferred to the PS where it is further accelerated to 26 GeV. It is in the PS that the particles are grouped into a train of bunches, each containing one hundred billion protons. Each bunch is about 1.2 meters long and they are separated by 7 meters. This separation is maintained all the way to collision in the LHC.

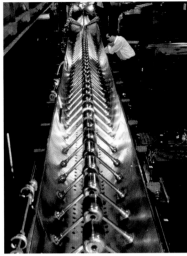

11. The source of all the protons. Hydrogen from the bottle in the background is ionized in the duoplasmatron source and protons are extracted through the nozzle on the right by applying a high voltage.

12. One of the accelerating tanks of the original CERN Linac with the top removed. The particles travel through a series of "drift tubes." An oscillating electric field in the gap between the drift tubes accelerates the particles. When the field is in the wrong direction, the particles are inside the drift tube and therefore shielded from the decelerating field.

13. The workhorse of CERN, the Proton Synchrotron, built in 1959.

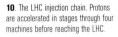

10. The LHC injection chain. Protons are accelerated in stages through four machines before reaching the LHC.

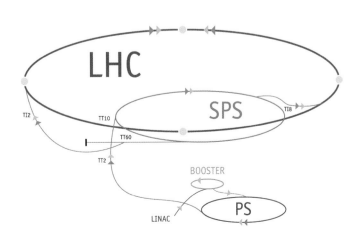

The Super Proton Synchrotron

At 26 GeV, the bunches are transferred into the next machine in the chain, the 1100-meter-radius Super Proton Synchrotron (SPS) built in 1976, where they are further accelerated up to 450 GeV and injected into the LHC, first in one ring and then into the other. When the two rings are filled, the magnetic field of the LHC is slowly ramped up and the beams are simultaneously accelerated by the radio frequency system (Chap. 4.5) which keeps them in the center of the vacuum chamber as the magnetic field rises. After about 20 minutes the beams reach the nominal collision energy of 7 TeV. They are then steered into collision in each of the four detectors.

At nominal intensity, there will be millions of collisions per second. However, in view of the enormous number of protons in each bunch, the intensity only decays very slowly. Typically the beams will remain in collision for about 10 hours after which any remaining beam is safely dumped onto an absorber block and the machine is brought back to its injection energy where the whole filling and acceleration cycle is repeated.

The injection of lead ions for the heavy ion program is slightly more complicated. It needs a special source and a Linac capable of accelerating lead ions. In order to get sufficient intensity, it also needs an accumulator ring where several Linac pulses can be accumulated before transferring the beam to the PS. The rest of the path through the PS and SPS to the LHC is similar to that of protons.

The design of the LHC

The fact that the LHC was to be constructed at CERN making the maximum possible use of existing infrastructure to reduce cost imposed a number of strong constraints on the technical choices to be made.

The first of these was the 27-km circumference of the LEP tunnel. The maximum energy attainable in a circular machine depends on the product of the bending radius in the dipole magnets and the maximum field strength attainable. Since the bending radius is constrained by the geometry of the tunnel, the magnetic field should be as high as possible. The field required to achieve the design energy of 7 TeV, is 8.3 tesla, about 60% higher than that achieved in previous machines. This pushed the design of superconducting magnets and their associated cooling systems to a new frontier.

The next constraint was the small (3.8 m) tunnel diameter. It must not be forgotten that the LHC is (just like the ISR) not one but two machines. A superconducting magnet occupies a considerable amount of space. To keep it cold, it must be inserted into an evacuated vacuum vessel called a cryostat and well insulated from external sources of heat. Due to the small transverse size of the tunnel, it would have been impossible to fit two independent rings, like in the ISR, into the space. Instead, a novel and elegant design with the two rings separated by only 19 cm inside a common yoke and cryostat was developed. This was not

14. The LHC.

15. A cross-section of the two-in-one LHC bending magnet. The two rings are concentrated inside a single vacuum vessel to save space (and money) see Fig. 10 of Chap. 4 for a more detailed view.

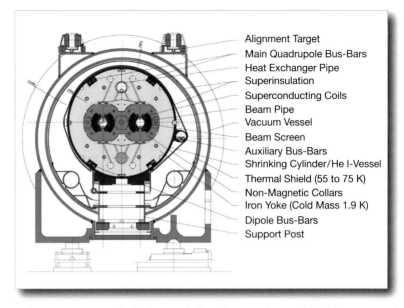

Alignment Target
Main Quadrupole Bus-Bars
Heat Exchanger Pipe
Superinsulation
Superconducting Coils
Beam Pipe
Vacuum Vessel
Beam Screen
Auxiliary Bus-Bars
Shrinking Cylinder/He I-Vessel
Thermal Shield (55 to 75 K)
Non-Magnetic Collars
Iron Yoke (Cold Mass 1.9 K)
Dipole Bus-Bars
Support Post

only necessary on technical grounds but also saved a considerable amount of money, some 20% of the total project cost.

Finally, the re-use of the existing injector chain governed the maximum energy at which beams could be injected into the LHC.

Machine layout

In parallel with the approval of the LHC machine, proposals for the experimental program were being examined by the LHC Experiments Committee (LHCC) whose job it was to give advice to the CERN management and through it to Council. Unlike the machine, the detectors have considerable independence. Only 20% of their funding comes through CERN. The rest comes from collaborating institutes all around the globe. However, it is the responsibility of CERN to provide the infrastructure, including the caverns in which the experiments are housed. Eventually, the LHCC proposed approval of two large general purpose detectors, ATLAS and CMS, as well as two smaller more specialized detectors, ALICE for heavy-ion physics and LHCb for the study of matter-antimatter asymmetry.

Civil engineering
The first job was to decide where these detectors were to be located. The LHC ring is segmented into eight identical arcs joined by eight 500- m Long Straight Sections (LSS) labeled from 1 to 8. Four of these LSS (at Point 2, 4, 6 and 8) already contain experimental caverns in which the four LEP detectors were located. These caverns are big enough to house the two smaller experiments. ATLAS and CMS required much bigger caverns, where excavation had to start while LEP was still running, the four even points therefore being excluded. Point 3 lies in a very inhospitable location deep under the Jura mountains and for various reasons, Point 7 could also be excluded. There remained Point 1, conveniently situated opposite the CERN main campus and diametrically opposite to Point 5, the most remote of all. Needless to say, there was considerable pressure from both ATLAS and CMS collaborations to get the more convenient Point 1. In the end, geology prevailed. Sample borings showed

16. Excavation of ATLAS. The cavern is the largest ever built in the type of rock encountered in the Geneva basin.

17. The inauguration of the ATLAS cavern on 4 June 2003 in the presence of the President of the Swiss Confederation, Pascal Couchepin (fourth from the left).

that Point 1 was much better suited for the larger cavern required for ATLAS. CMS was allocated Point 5, which at least had the advantage of being spared an endless flux of visitors. ALICE re-used the large electromagnet magnet of one of the old LEP experiments at Point 2 and LHCb was assigned the cavern at Point 8.

The excavation of the large caverns at Points 1 and 5 posed different problems. At Point 1, the cavern is the largest ever excavated in such ground conditions. The work also had to continue whilst the LEP machine was still operating. At Point 5, although the exploratory borings showed that there was a lot of ground water to be traversed when sinking the shaft, the speed of water flow took us by surprise. Extensive ground freezing was necessary to produce an ice wall around the shaft excavation.

An additional complication at Point 5 was that during the preparation of the worksite, the foundations of an ancient Roman farm (4th century A.D) were discovered. Work was immediately stopped so that the mandatory archeological investigation could be made. Articles of jewelry and coins minted in London, Lyon and Ostia, the ancient harbor city 35 km south-west of Rome, were found. The coins minted in London were dated 309-312 A.D, proving that in those days, the UK was part of the single European currency zone! But in those days, they didn't have a choice. One striking feature easily seen from the air (Fig. 18), is the precise alignment of the villa with respect to the boundaries of the present-day fields. This is evidence that the "cadastre", or land registry of today, is derived from the time of the Roman occupation.

A third civil engineering work package was the construction of two 2.6-km-long tunnels connecting the SPS to the LHC and the two beam dump tunnels and caverns.

Machine utilities

It takes more than just magnets to make a particle accelerator. Once the four straight sections were allocated to the detectors, the other four could be assigned to the essential machine utilities.

Figure 22 shows a schematic layout of the LHC ring. The two beams cross from one ring to the other at the four collision Points 1, 2, 5 and 8; elsewhere, they travel in separate vacuum chambers. They are transported from the SPS through two 2.6-km-long tunnels. Due to the orientation of the SPS with respect to the LHC, these tunnels join the LHC ring near Points 2 and 8. It was therefore necessary to integrate the injection systems for the two beams into the straight sections of the ALICE and LHCb detectors.

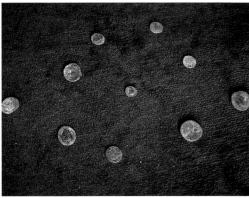

18. Aerial view of Point 5 in 1998. In the bottom of the picture are the original buildings from LEP. The foundations of a Roman farm from the 4th century can be seen top-center. Note how its walls are aligned perfectly with the boundaries of the surrounding fields.

19. Roman coins found during archeological excavations at Point 5. The larger coins are from the Emperor Maxence minted in Ostia between 309 and 312 AD. The smaller coins are from the Emperor Constantin minted in London and Lyon between 313 and 315.

20. An underground river made the excavation of the shaft of the CMS cavern very difficult. A ring of pipes carrying liquid nitrogen was used to form a wall of ice inside which the shaft was excavated and lined with concrete.

21. On February 1, 2005, the CMS cavern was inaugurated.

22. Schematic machine layout.

23. When protons are accelerated, the beam size becomes smaller. The collimators restrict the aperture to 12 mm at the injection energy but at 7 TeV they are closed to restrict the aperture to the size of the Iberian Peninsula on a one-euro coin.

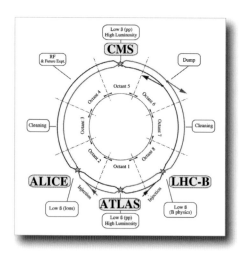

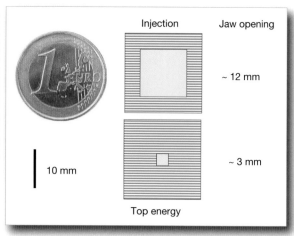

Clockwise from Point 2, the long straight section at Point 3 lies deep below the Jura mountains. It contains no experimental cavern from the LEP days and moreover, it is known from the experience of excavating the LEP tunnel that the geological conditions in this region are very difficult. Cracks and fissures in the rock allow water to percolate from the very top of the mountain, more than 1000 m high, producing a large static water pressure. In view of this it was decided that no additional civil engineering for tunnel enlargement would be allowed in this region. It was therefore assigned to one of the two collimation systems, which could be fitted into the existing tunnel.

Collimation is essential in a collider. As the beams are stored for many hours, a halo of particles slowly builds up around the core, mainly due to nonlinearities in the magnetic field or by the interaction of one beam with the other (in a lepton machine this halo would be damped by synchrotron radiation). If it were left uncontrolled, eventually particles would hit the vacuum chamber wall, producing unacceptable background in the detectors and risking a *quench* (a transition from the superconducting state due to the accompanying temperature rise) in some of the magnets. Collimators are specially designed motorized blocks that can be driven into the machine aperture to "clean" the beam by removing the halo locally. The collimators constitute the primary aperture restriction in the machine. When they are at their operating positions, the machine aperture is equivalent to the size of the Iberian Peninsula on a one euro coin (Fig. 23)!

Point 4 is assigned to the all-important Radio Frequency (RF) acceleration system. Acceleration is obtained by a longitudinal oscillating electric field at a frequency of 400 Megahertz (MHz) in a set of resonant cavities. The electric field in the cavities is very high, in excess of 5 million volts per meter. Once again, superconductivity comes to the rescue. The cavities are made of copper but there is a thin film of Niobium deposited on the inside surface. When cooled with liquid helium, this film becomes superconducting, enabling currents to flow in the cavity walls without loss.

Each revolution, the beam is given a small increase in energy as long as the field is pointing in the right direction. To achieve this, the frequency of the RF must be a precise harmonic of the revolution frequency so that each time a particle comes around, the field is pointing in the same direction. As the energy slowly increases, the magnetic field must also rise to keep the beams in the center of the vacuum chamber since the magnetic field required to bend a particle on a constant radius is proportional to its energy. The RF system needs considerable infrastructure and profits fully from the space available in the old LEP cavern at Point 4.

At 7 TeV with nominal intensity, the stored energy in one of the beams is 350 MJ, equivalent to more than 80 kg of TNT. If, for any reason this beam is lost in an uncontrolled way, it can do considerable damage to machine components, resulting in months of down-time. It is therefore essential to have a system that can reliably extract the beams very quickly and deposit them on special absorber blocks. This "beam-dump" system is located at Point 6 (and described in detail in Chap. 4). A set of special magnets can be pulsed very rapidly to kick the whole beam out of the machine in a single turn. The extracted beams are transported 700 m in an evacuated pipe and deposited on absorber blocks specially designed to take the enormous power.

The beam dump can be triggered by many sources, for instance if an excessive beam loss on the collimators is detected or if a critical power supply fails. It is also used routinely during operation; when the intensity in the beams

24. The superconducting
radio-frequency cavities at Point 4.

falls too low the beams are "dumped" by the operators in order to prepare the machine for the next filling cycle.

Finally, Point 7, like Point 3, contains a second collimation system.

The long straight sections each side of the four detectors house the equipment needed to bring the beams together into a single vacuum chamber and to focus them to a small spot with a radius of about 30 microns at the collision points inside the detectors.

International collaboration

The building of detectors has, for many years, been the subject of international collaboration. CERN, as the Host Institute, is generally responsible for providing the infrastructure including the experimental caverns and all the services needed to run the detectors but plays a minor role in the construction of the detectors themselves. These are managed by collaboration boards which divide the work of building the different components among many collaborating Institutes and Universities all around the world. This model has worked surprisingly well over the years considering the fairly loose management structure which has no direct hierarchical control over individual Institutes. The common cause seems to be a very strong motivator in keeping individual Institutes on track. This is the model that has been used for all four large detectors of the LHC.

Each LHC experiment will produce about 10 petabytes of data per year (1 PB = 1,000,000 GB). This corresponds to about 20 million DVD's. The analysis of this data requires enormous computing power, equivalent to about 100,000 of today's fastest PC processers. The collaborating Institutes are spread all over the world and need access to data locally. In order to address these needs an enormous amount of effort has been put into developing the LHC Computing Grid. The Grid infrastructure ties together hundreds of thousands of processers all around the world. Scientists can access data through the Grid freely from their home Institutes without having to know where it is stored or where it is being processed. More details about the Grid are provided in Chapter 5.6.

Although detector construction has a long tradition of broad international collaboration, machine construction has generally been the sole responsibility of the host laboratory. However, in the case of the LHC, it was necessary to find external collaborators to relieve the financial burden and the load on the CERN staff. External contributions, at least from the larger countries were also needed to justify access to the LHC experimental program. In the end, it has been a great success. Machine components have been built in Canada, India, Japan, Russia and the USA in a similar way to the construction of the detectors. Extra manpower has also been provided by several of CERN's Member States for the design of components and to help in the enormous effort of commissioning the machine hardware. Without this help, the LHC would not have been possible.

An invitation to explore the LHC

The construction of the LHC and its detectors has been a mammoth task. Right across the board, technologies have been pushed to their limits. In the following chapters, you will learn more about the construction of the machine and the four detectors from the people who built them. You will also learn how the challenge of finding the enormous computing power required to analyze the vast amount of data coming from the LHC detectors will be met.

The LHC is truly a tribute to human ingenuity and of international collaboration on a massive scale.

2.0

THE FUNDAMENTAL PHYSICS
BEHIND THE LHC

John Ellis

The fundamental scientific purpose of the LHC is to explore the inner structure of matter and the forces that govern its behavior, and thereby understand better the present content of the Universe and its evolution since the Big Bang, and possibly into the future. The unparalleled high energy of the LHC, which is designed to be 7 TeV per proton in each colliding beam, and its enormous collision rate, which is planned to attain about a billion collisions per second, will enable the LHC to examine rare processes occurring at very small distances inside matter. It will be a microscope able to explore the inner structure of matter on scales an order of magnitude smaller than any previous collider. The energies involved in these proton-proton collisions will be similar to those in particle collisions in the first trillionth of a second of the history of the Universe. By studying these processes in the laboratory, the LHC experiments will, in a sense, be looking further back into time than is possible with any telescope.

Although the collision energies achieved with the LHC will dwarf those achieved with previous accelerators, they are nevertheless exceeded regularly by the highest-energy cosmic rays striking the Earth. However, these collisions are very difficult to observe, and their energies, locations and timings cannot be controlled, unlike in the LHC. This control will enable the LHC to study in detail the processes occurring naturally in these cosmic processes, and unravel the underlying physics.

At the energies and distances explored with previous accelerators, matter is described successfully by a very precise theory called the Standard Model of particle physics, which is described in more detail in the next Section. According to the Standard Model, matter is composed of fundamental constituents, fermions called quarks and leptons, and the forces between them are carried by other particles, the fundamental bosons. Experiments at lower-energy accelerators agree very well with the predictions of the Standard Model.

However, the Standard Model is incomplete, and raises – but leaves unanswered – many fundamental questions. These include the origin of particle masses, the small difference between matter and antimatter, and the possible unification of the fundamental interactions. Moreover, the Standard Model has no explanation for some of the basic puzzles of cosmology, such as the origin of matter, and the natures of dark matter and dark energy.

As discussed in this Chapter, there are high hopes that the LHC will make breakthroughs in resolving at least some of these basic issues in physics beyond the Standard Model and in cosmology. In particular, detailed simulations show that the LHC will be able to discover the Higgs boson, the particle that is postulated within the Standard Model to give mass to the others, or else to discover what physics beyond the Standard Model replaces it. In many extensions of the Standard Model, there are also good hopes that experiments at the LHC will observe particles of dark matter. If we are lucky, studies at the LHC of the differences between matter and antimatter may cast light on the mechanism whereby matter came to dominate over antimatter in the early Universe. The LHC may also provide some circumstantial evidence for theories of all the particle interactions, e.g., by revealing supersymmetry and/or extra dimensions of space, both of which are essential ingredients in string theory.

The LHC is therefore poised to make many breakthroughs in the fundamental physics of matter and the Universe.

The Standard Model of particle physics

The Standard Model is based on the discoveries made in the early twentieth century that ordinary matter consists of electrons and nuclei, the latter being assemblages of protons and neutrons. It was discovered in the latter part of the twentieth century that protons and neutrons are themselves composite objects, made out of (as far as we know today) fundamental particles called quarks. Experiments with cosmic rays and accelerators have shown that there are six species of quark. Together with the electron and two similar particles, the muon and tau lepton, and their neutrino partners, these comprise the roster of the basic elementary particles that make up the Standard Model, displayed in Figure 1.

There are four fundamental forces acting on these elementary particles, illustrated in Figure 2. Two of them are very familiar: electromagnetism, which is the dominant force at the atomic and molecular scales, and gravity, which is relatively unimportant at the distances and energies probed at accelerators, though very important at macroscopic scales. In addition, there are two short-range forces that are important at the elementary-particle level, namely the strong and weak nuclear forces. The first of these holds nuclei together, and the latter causes, among other phenomena, the beta decays of neutrons, heavier quarks and leptons.

The Standard Model places these fundamental particles and the forces between them on a solid theoretical basis that enables very detailed calculations to be made for the strong, weak and electromagnetic forces between the

1. The fundamental building blocks of matter in the Standard Model.

2. The four fundamental forces: gravity, electromagnetism, and the weak and strong nuclear forces.

fundamental particles. Each of these forces is described classically by a field extending through space, and is represented at the quantum level by the exchange of some species of particle. In the case of electromagnetism, the field quantum is the photon, the corresponding quanta of the weak nuclear forces are the massive W and Z vector bosons discovered at CERN in 1983, and the quantum of the strong nuclear forces is the gluon, discovered at the Deutsches Elektronen Synchrotron (DESY) in 1979. The corresponding quantum of the gravitational field is the hypothetical graviton, which has not been observed. (Nor indeed have classical gravitational waves, though indirect effects of their emission on the motions of pulsars have been seen.)

The great merit of the Standard Model is that it unifies the descriptions of the strong, weak and electromagnetic forces, and enables accurate calculations of their effects to be made. In particular, one can calculate reliably quantum corrections to processes involving these forces, as was first shown by Gerardus 't Hooft and Martinus Veltman in 1970. Subsequent theoretical calculations made using the Standard Model agree very well with data taken at accelerators that have operated at lower energies than the LHC, such as CERN's LEP electron-positron accelerator in the 1990s and more recently the proton-antiproton collider at Fermilab. In particular, data taken at LEP agreed with the Standard Model at the per-mille level, this success enabled the mass of the top quark to be predicted successfully before its discovery at Fermilab. Subsequently, 't Hooft and Veltman shared the 1996 Nobel Physics Prize for making possible this success of the Standard Model.

The Holy Grail of particle physics

However, these theoretical calculations are valid only with a missing ingredient that has not yet been observed, namely the so-called Higgs boson. Without this missing ingredient, the calculations would yield incomprehensible infinite results. The agreement of the data with the calculations implies not only that the Higgs boson (or something equivalent) must exist, but also suggests that its mass should be well within the reach of the LHC.

Why should this Higgs boson exist, and are there any alternatives? In the underlying equations of the Standard Model, none of the elementary particles appear to have masses. However, in the real world only the photon and gluon, the carriers of the electromagnetic and strong nuclear interactions, respectively, are massless. All the other elementary particles are massive, with the W and Z vector bosons and the top quark weighing as much as decent-sized nuclei. The underlying symmetry between the different particles of the Standard Model must be broken, so that some may acquire masses while others remain massless.

There are two ways to break the underlying symmetry of the Standard Model. The preferred way is to respect the symmetry in the fundamental equations, but look for an asymmetric solution of them, much as the reader and writer are lopsided solutions of physical equations that have no preferred orientation. In the Standard Model, the breaking of the symmetry is thought to be present already in the lowest-energy state, namely the so-called vacuum. This 'spontaneous' symmetry breaking is ascribed to a field that permeates all space, taking a specific value that can be calculated from the underlying equations, but with a random orientation in the internal 'space' of particles that the underlying symmetry.

This idea was first exploited in condensed-matter physics by Phil Anderson and others, and related ideas were introduced into particle physics by Yoichiro Nambu, winner of half of the 2008 Nobel Physics Prize. The idea that this mechanism might be used to provide masses for vector bosons was suggested by Robert Brout and Francois Englert, and independently by Peter Higgs. As incorporated into the Standard Model, their mechanism requires some elementary particles to remain massless, such as the photon, but gives masses to others, such as the electron, the quarks and the W and Z vector bosons, in proportion to their couplings to this vacuum Brout-Englert-Higgs (BEH) field.

An analogy may help the reader to visualize how this BEH mechanism works. Imagine a flat, featureless snowfield, extending uniformly throughout space. Now consider a skier crossing it: (s)he glides rapidly across the snow without sinking into it, much as a massless particle such as the photon (which does not interact with the BEH field) always travels very fast, namely at the speed of light. Consider next a snowshoer crossing the snowfield: (s)he sinks a bit into the snow and moves more slowly than the skier, rather like a massive particle that interacts with the BEH field, which thereby acquires a mass and always travels slower than light. Finally, consider a hiker: (s)he sinks deeply into the snow and moves very slowly, just like a particle with large mass that interacts strongly with the BEH field.

Just as the electromagnetic field has a quantum particle associated with it, namely the photon, this vacuum BEH field would also have an associated quantum particle: a snowflake, if you like! This is the long-awaited Higgs boson, so called because Higgs was the first to notice that its existence is an inevitable consequence of the theory. It has now become the Holy Grail of particle physics, that the large LHC experiments, ATLAS and CMS, will be seeking. Unfortunately, the Standard Model makes no *a priori* prediction what the mass of the Higgs boson might be.

Experiments at LEP at one time found a hint for its existence, but finally these searches were unsuccessful, and told us only that it must weigh at least 114 GeV (about 120 times the mass of the proton). If its mass is less than about 200 GeV, the Fermilab proton-antiproton collider may be able to find some hint of the existence of the Higgs boson before the LHC comes into operation. So far, however, the Tevatron collider experiments have found no evidence for it, and even exclude the possibility that it weighs between 160 and 170 GeV. Just as LEP measurements were used to predict successfully the mass of the top quark, they and the more recent Tevatron measurements of the

3. Simulation of the production and decay of a Higgs boson into an electron and positron (which deposit their energies on the right side, seen as two blue towers) and a muon-antimuon pair (seen as the yellow lines on the left side).

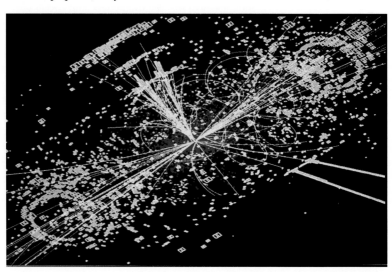

masses of the W vector boson and the top quark suggest (if the Standard Model is right) that the Higgs boson weighs less than about 140 GeV.

The large LHC experiments, ATLAS and CMS, will be looking for the Higgs boson in several different ways. The Higgs boson is predicted to be unstable and decay into other particles, such as photons, bottom quarks, tau leptons, W or Z bosons. One simulation of Higgs boson production and decay at the LHC is shown in Figure 3. It may well be necessary to combine several different decay modes in order to uncover a convincing signal. However, the LHC experiments should be able to find the Higgs boson even if it weighs as much as 1 TeV, and there are good expectations that a Higgs boson could be found during the first couple of years of LHC operation at the design energy and luminosity. Its discovery would set the seal on the success of the Standard Model.

If not the Higgs Boson, then what?

With the impending confirmation or disproof of the Brout-Englert-Higgs idea, via the discovery (or proof of non-existence) of the Higgs boson, many theorists are getting cold feet, and staking their claims to alternative scenarios that go beyond the Standard Model.

One popular suggestion has been that the Higgs boson might not be an "elementary" particle in the same sense as the quarks, leptons and photon, but might be composed of more elementary constituents. This scenario would be analogous to the Bardeen-Cooper-Schrieffer (BCS) theory of superconductivity, in which the photon acquires an effective mass by interacting with Cooper pairs of electrons that are bound together. The analogy in particle physics would be to replace the elementary BEH field by a composite object made out of some new, as-yet-unknown matter particles held together by superstrong, as-yet-unknown forces.

It seems rather difficult to reconcile this BCS-inspired composite alternative with the accurate data available from LEP and the Tevatron, but some enthusiasts are still pursuing this possibility. Such a scenario would predict a cornucopia of new, heavy particles, the excited bound states of the new, as-yet-unknown matter particles, which would certainly be exciting for the LHC!

The most radical alternative to the Higgs hypothesis is to exploit a different way of breaking the Standard Model symmetry, namely to postulate that, while the underlying equations are indeed symmetric, their solution is subject to boundary conditions that break the symmetry. What boundary, since space is apparently infinite, or at least very large compared to the scale of particle physics? The answer is that there may be additional, very small dimensions of space with edges where the symmetry between the different particles may be broken.

Such models would have no Higgs boson, which sounds like bad news for the LHC. However, in such a "Higgsless" model, the W and Z bosons would tend to interact strongly together at high energies, an effect that could in principle be seen at the LHC. However, such strong W and Z interactions are difficult to reconcile with the present low-energy data that seem to require a relatively light Higgs boson.

Physicists are amusing themselves discussing what would be the maximum credible disaster scenario for the LHC: discover a Higgs boson with the properties predicted in the Standard Model, or discover that there is no Higgs boson? The former would be a vindication of theory, but would teach us little new, while the latter would reopen the entire basis of the Standard Model.

Hence the absence of a Higgs boson would be exciting for particle physicists, though it might not be so easy to explain to the politicians who have funded the LHC with the discovery of the Higgs boson as a central motivation. However, whichever option Nature chooses, the good news is that the LHC will provide us with a clear-cut experimental answer, putting paid to all speculations!

If the Higgs Boson, then what?

Resolving the Higgs question will set the seal on the Standard Model, or perhaps put a nail in its coffin. However, as mentioned in the opening paragraph, there are plenty of other reasons to expect new physics beyond the Standard Model. Specifically, there are good reasons to expect new physics at the TeV energy scale, within reach of experiments at the LHC. Indeed, some might consider these to be the primary motivations for the leap into the unknown that the LHC represents, even more than the iconic Higgs boson.

For example, it is generally thought that the elementary Higgs boson of the Standard Model probably could not exist in isolation. Specifically, within the Standard Model one can calculate the overall energy of the vacuum as a function of the magnitude of the BEH field. There are two potential disasters, the Scylla and Charybdis of Higgs physics! For example, if the mass of the Higgs boson is too large, the energy of the BEH field grows, and there is nothing in the Standard Model to prevent it from becoming infinitely large: the Higgs Scylla. Presumably some new physics beyond the Standard Model comes in to damp down the over-enthusiastic BEH field energy, but what? Alternatively, if the mass of the Higgs boson is too small, the energy of the BEH field actually turns negative, and our present vacuum would be unstable, and would eventually decay into this Higgs Charybdis.

The existences of the Scylla and Charybdis of Higgs physics require us to refine our concept of the maximal credible disaster scenario for the LHC: if the Higgs it discovers is either very heavy or very light, there must be some new physics beyond the Standard Model that might also appear at the LHC. The true disaster would be if the Higgs boson weighs about 180 GeV, in which case there would apparently be no need for any physics beyond it. Reassuringly, the present data seem to favor a relatively light Higgs boson, in which case we are doomed to spiral into the Higgs Charybdis unless some new physics intervenes. The new physics required to avert this theoretical disaster would have to look a lot like supersymmetry, the cure-all theory that also has many other motivations, as we now discuss.

The first of these is provided by the difficulties that arise when one calculates quantum corrections to the mass of the Higgs boson. Not only are these corrections infinite in the Standard Model, but, if the usual procedure is adopted of controlling them by cutting the theory off at some high energy or short distance, the net result depends on the square of the cutoff scale. This implies that, if the Standard Model is embedded in some more complete theory that kicks in at high energy, such as a grand unified theory of the particle interactions or a quantum theory of gravity, the mass of the Higgs boson would be very sensitive to the details of this high-energy theory. This would make it difficult to understand why the Higgs boson is expected to have a (relatively) low mass and, by extension, why the scale of the weak interactions is so much smaller than that of unification or quantum gravity.

One might be tempted simply to wish away this 'hierarchy problem', by postulating that the underlying parameters of the theory are tuned very finely, so that the net value of the Higgs boson mass obtained after adding in the quantum corrections is unnaturally small as the result of some sneaky cancellation. This is probably mathematically possible, but it seems physically unreasonable, even unnatural.

Surely it would be more satisfactory either to abolish the extreme sensitivity to the quantum corrections, or to cancel them in some systematic manner. Indeed, this used to be one of the primary motivations for believing that the Higgs boson might be composite. In that case, the Higgs boson would have a finite size, which would cut the pesky quantum corrections off at some relatively low scale. In that case, as already mentioned, the LHC might discover a cornucopia of new particles with masses around this cutoff scale, which should be near a TeV, or at least the interactions of the W and Z bosons would be modified in an observable way.

Why Supersymmetry?

The alternative is to cancel the quantum corrections systematically, which is where supersymmetry could come in. Supersymmetry is a very elegant theory that would pair up fermions, such as the quarks and leptons that make up ordinary matter, with bosons, such as the photon, gluon, W and Z that carry forces between the matter particles, or even the Higgs boson itself. Supersymmetry also seems be essential for making a consistent quantum theory of gravity based on string theory (of which more later). However, these elegant arguments give no clue what energies would be required to observe supersymmetry in nature.

The first practical argument that supersymmetry might appear near the TeV scale was provided by the hierarchy problem: in a supersymmetric theory, since for each fermion there is a corresponding boson, the quantum corrections due to the effects of fermions and bosons cancel each other systematically, and the large hierarchy of mass scales in physics no longer appears unnatural. The residual quantum corrections to the mass of the Higgs boson would be small if differences in masses between supersymmetric partner particles would be less than about 1 TeV. Why this should be the case is another question, but at least the naturalness aspect of the hierarchy problem is solved.

The only snag is that the fermions and bosons of the Standard Model do not pair up with each other in a neat supersymmetric manner. Therefore, this theory would require each of the Standard Model particles to be accompanied by an as yet unseen supersymmetric partner. This hypothesis may seem profligate, but at least it predicts a 'scornucopia' of new particles that should weigh less than about a TeV, and hence could be discovered by the LHC.

In the wake of this hierarchy argument, at least five other reasons have

4. The cosmic 'pizza' has a strange combination of 'toppings': the density of cold dark matter is several times larger than that of conventional visible matter. CMB stands for *cosmic microwave background*.

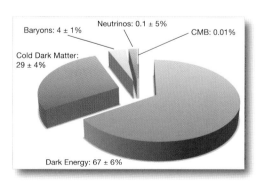

Neutrinos: 0.1 ± 5%

Baryons: 4 ± 1%

CMB: 0.01%

Cold Dark Matter: 29 ± 4%

Dark Energy: 67 ± 6%

5. When extrapolating the strengths of the Standard Model couplings to high energies, they do not unify if only the Standard Model particles are included (left panel), but could unify if supersymmetric particles are included (right panel).

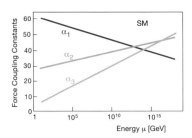
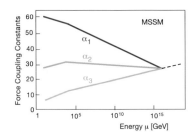

surfaced for thinking that supersymmetric particles might weigh about a TeV.

Historically, the first was that, in many supersymmetric models, the lightest supersymmetric particle (LSP) would be an ideal candidate for dark matter. Astrophysicists and cosmologists tell us that some 80% of the matter in the Universe is made of some invisible stuff of a different nature from the visible matter, as seen in Figure 4. Explaining dark matter requires some new physics beyond the Standard Model, and supersymmetry is an ideal candidate. This is because the LSP is stable if a suitable combination of baryon and lepton numbers is conserved. This is indeed what happens in the minimal supersymmetric extension of the Standard Model, as well as in simple models of grand unification and neutrino masses. In this case, the stable LSPs would be left over as relics from very early in the Big Bang, and calculations of their abundance yield a density of dark matter in the range favoured by astrophysics and cosmology if the LSP weighs at most a few hundred GeV, probably putting it within reach of the LHC.

The second motivation for low-energy supersymmetry was the observation that they would facilitate the unification of the strong, weak and electromagnetic forces in some simple grand unified theory. Starting from the strengths of the fundamental strong, weak and electromagnetic forces measured at LEP, it was possible to extrapolate them to much higher energies. In a grand unified theory, one would expect the coupling strengths of the three basic forces of the Standard Model to become equal at some high energy. This is not the case in the Standard Model, as seen in the left panel of Figure 5. However, if one adds to the particles of the Standard Model the effects of their hypothetical supersymmetric partners, and assumes that their masses are about a TeV, there is a high energy (around 10^{16} GeV) where all the Standard-Model forces have equal strengths, as seen in the right panel of Figure 2.5, and grand unification becomes thinkable.

A third argument is that a theory with low-energy supersymmetry would predict that the Higgs boson weighs less than about 150 GeV. As discussed above, this is precisely the range that is favored indirectly by the present data from LEP and the Tevatron, as shown in Figure 6.

6. Combining the negative results of direct searches for the Higgs boson with indirect indications from high-precision electroweak data, the most likely value of the Higgs mass is about 120 GeV. The three upper limits have theoretical underpinnings that might prove incorrect; it is thus important to have the LHC's energy range in order to search at much higher values as well.

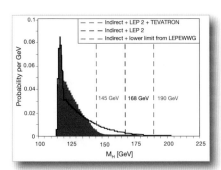

A fourth motivation for low-energy supersymmetry is the desire to avoid the Higgs Charybdis of an unstable vacuum, as also discussed above.

Finally, mention should be made of the one low-energy experiment that does not agree *prima facie* with the Standard Model: the measurement of the magnetic moment of the muon. The corresponding measurement for the electron agrees with the Standard Model to many decimal places, and is the most accurate verification of quantum electrodynamics. However, the experimental agreement with the Standard Model calculation is not so good for the muon. The situation is not clear-cut, because the Standard Model calculation requires input from other experiments, specifically either low-energy positron-electron (e^+e^-) annihilation or the decays of tau leptons into hadrons. Using e^+e^- data generally yields a significant discrepancy with the experimental value of the magnetic moment of the muon, whereas the discrepancy is not significant if tau decay data are used. Therefore, it is not clear whether new physics beyond the Standard Model is required at all. However, if it is needed, supersymmetry would fit the bill perfectly.

So, there are several reasons for hoping to find supersymmetry at the LHC. However, to paraphrase Richard Feynman, if you had one good reason you would not need to mention five. Nevertheless, I consider supersymmetry to be the best motivated extension of the Standard Model.

Supersymmetry at the LHC?

Supersymmetry could be a bonanza for the LHC, with many different types of supersymmetric particle available to be discovered. In many models, the LHC would produce pairs of strongly-interacting sparticles, either *gluinos* (the supersymmetric partners of the gluons) or *squarks* (the supersymmetric partners of the quarks). These would subsequently decay rapidly via various intermediate supersymmetric particles, such as the supersymmetric partners of the W, Z and Higgs bosons, or *sleptons* (the supersymmetric partners of leptons), yielding finally a pair of LSPs that would carry energy away invisibly. In this scenario, the key experimental signature would be an imbalance in the transverse momenta carried by the visible particles, as seen in Figure 7.

Many such scenarios have been simulated in the ATLAS and CMS detectors, with the conclusion that supersymmetry could be discovered in a large part of the parameter space of supersymmetric models. In favorable cases, the masses of several intermediate particles produced in the cascades could be reconstructed. With a bit of luck, one might learn enough about the supersymmetric spectrum to be able to calculate what the supersymmetric dark matter density should be, so as to compare the result with the astrophysical estimates. It would truly be a tremendous connection between microphysics and

7. Simulation of the production of a pair of heavy supersymmetric particles at the LHC, followed by their decays into invisible dark matter particles, leaving an apparent imbalance in the tranverse momentum.

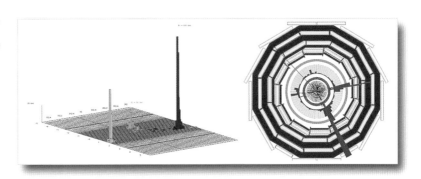

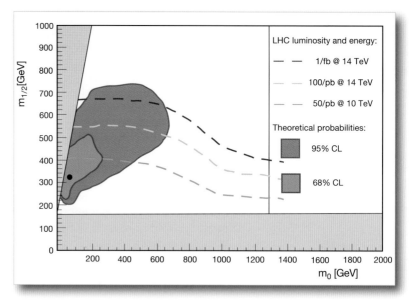

macrophysics if one was able to verify the nature of the cosmological dark matter in this way.

Further cause for optimism is given by a global fit to experimental data, with the precise measurements of electroweak quantities, including the magnetic moment of the muon, as well as the cosmological estimate of the cold dark matter density. In some of the simplest supersymmetric models, the masses favored for the supersymmetric particles place them well within the range accessible to the LHC as shown in Figure 8. However, this encouraging conclusion is very dependent on the interpretation of the magnetic moment of the muon.

It is also possible that supersymmetry might have a rather different experimental signature at the LHC. The classic missing-energy signature would hold if the LSP is some combination of the supersymmetric partners of the W, Z, Higgs boson and photon, called a *neutralino*. Another possibility is that the LSP is the supersymmetric partner of the graviton, the *gravitino*. In this case, the next-to-lightest supersymmetric particle (NLSP) might well be metastable, decaying into the gravitino only via gravitational-strength interactions. There are various possibilities for this NLSP: for example, it might be a *sneutrino* (supersymmetric partner of a neutrino), or a charged *slepton* (supersymmetric partner of a charged lepton), or a stop *squark* (supersymmetric partner of a top quark). The sneutrino NLSP case would give a classic missing-transverse-momentum signature, but a charged slepton would have a distinctive signature as a metastable charged particle, and a stop squark would not only be metastable, but would would form bound states that changed charge as they passed through the detector.

These examples show that the experimental signature of supersymmetry may be very different from what would be expected in simple models. The LHC experiments must keep their eyes open for the unexpected!

If not Supersymmetry, then what?

Postulating a composite Higgs boson or supersymmetry are not the only strategies that have been proposed for dealing with the hierarchy problem. Another suggestion has been to postulate additional dimensions of space. Clearly, space is three-dimensional on the scales that we know it so far, but the idea that there might be additional dimensions curled up so small as to be invisible has been in the air since it was first proposed by Kaluza and Klein over 80 years ago. This idea has gained currency in recent years with the realization that string theory predicts the existence of extra dimensions of space.

According to string theory, elementary particles are not the idealized points of Euclidean geometry, but are instead objects extended in one dimension (the eponymous strings) or as membranes in more dimensions. In order for the quantum theory of strings to be consistent, they have to move in a space with more than the usual three dimensions. In the original formulations of strings without fundamental fermions, there were thought to be 25 spatial dimensions. However, supersymmetry seems to be (almost) essential for the consistency of string theory, in which case it was thought that 9 spatial dimensions were needed. Subsequently this was increased to 10 in the general formulation known as M theory.

Initially, it was thought that these extra space dimensions should be very small, curled up on scales that might be as small as the Planck length, namely around 10^{-33} cm. However, more recently it has been realized that at least some of these new dimensions could be much larger, possibly even large enough to have consequences observable at the LHC.

If the extra dimensions are curled up on a sufficiently large scale, ATLAS and CMS might be able to see excitations of Standard Model particles, or even the graviton, which appear when particles spiral around an extra dimension as they move through the usual three-dimensional space. The spectroscopy of these 'Kaluza-Klein excitations' might, in some extra-dimensional theories, be as rich as that of supersymmetry. If so, how to tell which cornucopia the LHC uncovers? There are significant differences in the relations between, for example, the masses of the partners of quarks and leptons in supersymmetric theories and those with large extra dimensions. Moreover, the spins of the Kaluza-Klein excitations would be the same as those of their Standard Model progenitors, whereas the spins of supersymmetric partners would be different. These underlying differences translate into characteristic differences in the spectra of decay products in the two classes of model, and into distinctive correlations between them. Studies indicate that ATLAS and CMS should be able to distinguish between real extra dimensions and their supersymmetric imitators.

It is amusing that, in some theories with extra dimensions, the lightest Kaluza-Klein particle (LKP) might be stable, rather like the LSP in supersymmetric models. In this case, the LKP would be another candidate for astrophysical dark matter. Thus, there is more than one way in which LHC physics beyond the Standard Model might explain the origin of dark matter. Fortunately, as discussed in the previous paragraph, tools seem to be available for distinguishing between them.

One of the most dramatic possibilities offered by speculations about extra dimensions is that gravity might become strong as it spreads into these extra dimensions, possibly at energies close to a TeV. In this case, according to some variants of string theory, microscopic black holes might be produced by the LHC. The theory that predicts these microscopic black holes also predicts that they would be very short-lived, decaying rapidly via *Hawking radiation*, as shown in Figure 9. Measurements of this radiation would offer a unique laboratory window on the mysteries of quantum gravity. The microscopic black holes would emit energetic photons, leptons, quarks and neutrinos, providing distinctive experimental signatures. In particular, the neutrinos would carry away even more invisible energy than the supersymmetric models discussed previously.

9. Simulation of the production of a microscopic black hole at the LHC, followed by its decay via Hawking radiation.

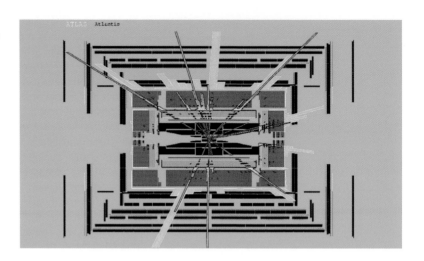

34

It has been suggested that microscopic black holes might instead be stable, and concern has been expressed that they might then gobble up the matter in the Earth and threaten humanity. Any such fears are groundless.

Independently from the argument that black holes should decay via Hawking radiation, elementary quantum physics would predict that any object produced by the collision of a pair of particles would be able to decay back into them. Since microscopic black holes would be produced, at the elementary-particle level, by the collisions of quarks and gluons, they should be able to decay back into the same particles.

Nevertheless, what if some microscopic black hole was stable? Recall that the Earth has been bathed in cosmic rays for billions of years, ever since it was formed, and some of the cosmic rays hit the Earth with higher energies than those the LHC will achieve. We calculate that the entire LHC experimental program has already been carried out on Earth some 100,000 times, and that every second it is being carried out over 10^{13} times by cosmic-ray collisions with stars throughout the visible Universe. If there was any danger from the paltry LHC collisions, the Earth, Sun and other stars would not exist.

There is an apparent loophole in this argument, namely that objects produced at the LHC would tend to be moving more slowly than the same objects produced in cosmic-ray collisions. However, detailed analysis shows this does not invalidate the basic argument. Electrically-charged black holes would stop inside the Earth or Sun even if they were produced by cosmic rays, and neutral black holes would stop inside neutron stars or white dwarfs. The fact that the Earth and Sun continue to exist, and that long-lived neutron stars and white dwarfs also exist, tells us that cosmic rays do not produce dangerous black holes capable of eating the planet. Therefore, neither will the lower-energy and less copious LHC collisions.

The Origin of the Matter in the Universe?

In the previous two sections we discussed ways in which the LHC could cast light on the dark matter in the Universe: is there any way in which it might help explain the origin of conventional matter? As was first pointed out by the Russian physicist Andrei Sakharov, the fact that matter and antimatter have slightly different properties, as discovered in the laboratory in the decays of K mesons, could enable particle physics to explain the origin of the matter in the

Universe. The other ingredients in the Sakharov mechanism are forces that generate matter, which are predicted in grand unified theories and even in the Standard Model, but have never been seen, and some non-equilibrium process in the early Universe. Combining these ingredients, the amount of matter could have come to exceed slightly the amount of antimatter during the evolution of the Universe, all the antimatter would subsequently have annihilated with matter particles to form radiation, and the small residual excess of matter is what we see in the Universe today.

The matter-antimatter differences seen in the laboratory, in both K and B meson decays, are in very good agreement with the mechanism of Makoto Kobayashi and Toshihide Maskawa, which operates within the Standard Model. Their mechanism requires six or more quarks, which seemed rather far-fetched when they proposed it in 1973, at a time when only three quarks were known. Their insight was rightfully recognized with the other half of the 2008 Nobel Physics Prize.

However, the Kobayashi-Maskawa mechanism in the Standard Model and the matter-antimatter differences seen in the laboratory so far would not have yielded enough matter. The differences are just not large enough, and a suitable non-equilibrium process in the early Universe has not been identified. The latest epoch at which such a process might have occurred was when the BEH mechanism switched on in the early Universe, when typical particle energies were about 100 GeV or so. However, in the Standard Model the BEH mechanism would have switched on very smoothly, and the Universe would have stayed in thermal equilibrium.

The Sakharov mechanism could work at the 100 GeV energy scale if there were additional sources of matter-antimatter differences, and if there were additional light bosonic particles around that would make the BEH mechanism switch on more jerkily. Both of these features might be present in new physics at the TeV scale that would be accessible to the LHC. For example, supersymmetry allows many more possibilities for differences between the properties of matter and antimatter than are possible in the Standard Model, some of which might be able to explain the amount of matter in the Universe. Furthermore, some supersymmetric bosons might be relatively light, helping to make the evolution of the Universe jerkier.

This provides one of the motivations for the LHCb experiment, which is dedicated to probing the differences between matter and antimatter, notably looking for discrepancies with the Standard Model. In particular, LHCb has unique capabilities for probing the decays of mesons containing both bottom and strange quarks, the constituents of the B and K mesons that were probed in other experiments on matter-antimatter differences. There are also many other ways to explore the physics of matter and antimatter, and the ATLAS and CMS experiments will contribute, in particular, by searching for rare decays of mesons containing bottom quarks.

If they detect any new particles beyond the Standard Model at the TeV scale, the question will immediately arise whether this new physics distinguishes between matter and antimatter, and whether or not this new physics could explain the origin of the matter in the Universe. For example, if the Higgs boson is discovered at the LHC, are its couplings to matter and antimatter the same? If supersymmetry is discovered at the LHC, do sparticles and antisparticles behave in the same way? There are testable scenarios in which matter-antimatter differences in the Higgs or sparticle sector could be responsible for the origin of the matter in the Universe.

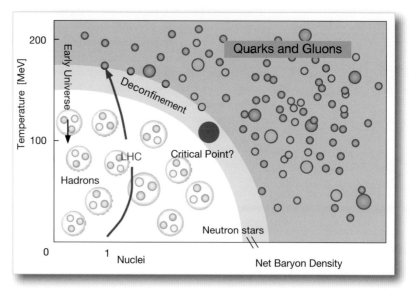

10. The phase diagram for nuclear matter. In the bottom left corner, quarks and gluons are confined into hadrons, whereas at high temperatures they are deconfined. Other accelerators explore the transition at non-zero net baryon density, whereas the LHC will explore physics at low net baryon density, similar to that in the early Universe.

Recreating the Primordial Plasma

In discussing the relations between LHC physics and the early Universe, we have been assuming implicitly the conventional Big Bang theory of the Universe. This was motivated initially by the present-day Hubble expansion, and subsequently by the observation of the cosmic microwave background radiation. This tells us that the Universe was once about a thousand times hotter and smaller than it is today, containing no atoms but only electrons, light ions and radiation. Currently, our earliest window into the Universe is the successful comparison between the light-element abundances observed by astrophysicists and those calculated by cosmologists using particle and nuclear physics within the standard Big Bang model. This tells us that the Universe was once about a billion times smaller and hotter than today, with a temperature around a billion degrees and typical particle energies around an MeV.

The LHC will help us look back further in time. By colliding relativistic heavy nuclei, it will create regions with effective temperatures around a trillion degrees, with typical particle energies approaching a GeV. At that epoch, it is thought that quarks and gluons acted as free particles, whereas afterwards they were confined inside hadrons. As the Universe cooled, the free quarks would have converted into hadrons, and the LHC will be able to recreate this quark-hadron transition on a microscopic scale, as illustrated in Figure 10.

Theorists have various ways to calculate the behavior of the primordial plasma and the nature of this transition, which can be tested by the ALICE experiment, in particular. One of the most exciting new theoretical approaches uses techniques borrowed from string theory, which relate the strongly-interacting plasma just above the transition temperature to a weakly- interacting gravitational system in five dimensions. If these theoretical calculations are consistent with the LHC measurements, our confidence in our description of the very early Universe will be reinforced, reducing the uncertainties in our speculations about the origins of visible and dark matter.

Into the Future

Even after the first full-energy collisions of the LHC take place, it will take some time for the accelerator to build up to its designed nominal collision rate. Much of the initial operation will serve to test our understanding of the Standard Model, for example by looking at the production of jets of strongly-interacting particles, and measuring the properties of the W and Z bosons and the top quark. Perhaps some surprises will emerge: maybe quarks will turn out to be composite, or perhaps there are additional W and Z bosons?

After this initial operation, one can hope that the LHC will start to provide key information on physics beyond the Standard Model, e.g., by discovering the Higgs boson, or by discovering other new particles such as those predicted by supersymmetry, if they are not very heavy.

Continued running of the LHC at the nominal luminosity would enable many properties of the Higgs boson to be verified, e.g., by providing measurements of its couplings to some other particles and checking whether they are proportional to their masses, and perhaps measuring its spin. This period should also enable the properties of any other new particles to be checked, e.g., whether their spins are the same or different from those of their Standard Model counterparts.

What might be possible using the LHC, beyond these planned phases of exploitation? One possibility is to add new components to the existing ATLAS and CMS detectors that could provide new ways to study the Higgs boson. Another possibility is that supersymmetric or other new particles might show up in unexpected ways. For example, in some supersymmetric scenarios there would be a metastable charged particle that would have quite distinctive experimental signatures, and it might be interesting to devise new detectors to explore this possibility in more detail.

It might also be possible to increase the LHC collision rate significantly beyond the nominal value. This possibility would be particularly interesting if, for example, the initial runs of the LHC discover new physics with a very low production rate, perhaps because it has a high-energy threshold. Some increase in the LHC collision rate would be possible by redesigning the collision points using new magnet technologies, but one would also like to replace at least some of CERN's lower-energy accelerators so as to feed more intense beams into the LHC. The first step in this program, the construction of a new low-energy linear accelerator, has already started, and other technical options for increasing the LHC collision rate are now being evaluated. Whether to proceed with them can be considered when the first experimental results from the initial LHC runs become available, some time around 2011.

Particle physics stands on the brink of a new era, with the LHC poised to make the first exploration of physics in the TeV energy range. There are good reasons to hope that the LHC will find new physics beyond the Standard Model, but no guarantees. The most one can say for now is that the LHC has the potential to revolutionize particle physics, and that in a few years' time we should know what course this revolution will take. Will there be a Higgs boson, or not? Will space reveal new properties at small distances, such as extra dimensions or supersymmetry? Will experiments at the LHC cast light on some fundamental cosmological questions, such as the origin of matter or the nature of dark matter? Whatever may be the answers to these questions, or whatever other surprises the LHC may provide, it will surely set the agenda for the next steps in fundamental physics.

The Fundamental Physics Behind the LHC

3.1

THE CONSTRUCTION OF THE LHC

**Lessons in Big Science
Management and Contracting**
Anders Unnervik

1. Delivery of a LHC cryogenic dipole magnet to the CERN.

The world's greatest and most complex scientific instrument ever conceived does not get built without organization, innovative procurement and careful oversight. In addition, the completion of an instrument on this scale, in the context of a vast international collaboration, with a nine-figure – yet highly restricted – budget, had to take the following constraints into account:
- significant technologies, production methods and instruments did not yet exist at the starting time of the project;
- spending had to take national interests into account and ensure a fair industrial return to the Member States;
- the procurement service had to navigate through risks of lowest-bidder economics and needed to figure out how to balance innovation and creativity with quality control and strict procurement procedures;
- a tight budget and project schedule required that the impact due to long lead times of essential components and tooling, business failures, cost overruns, disputes, etc. had to be minimized.

In the first part of this chapter, we will try to present the starting philosophy and discuss how this evolved under the reality of operations, illustrated by a number of specific examples. In particular, we will see how the management of the perceived risk from the supplier perspective was minimized, thus significantly reducing cost estimates and keeping the LHC very close to on-budget performance.

2. Atlas magnet toroid end cap being transported to ATLAS Point 1.

Supplier management

The total amount charged to the CERN budget for the LHC project is 4.332 billion Swiss francs. This amount includes CERN's expenses for R&D, machine construction, tests and pre-operation, LHC computing as well as CERN's contribution to the cost of the detectors. It is important to note that this figure does not include CERN's manpower costs (which total another 2.2 billion Swiss francs, cf. Table 1). In addition to all purchases made for the LHC and financed by CERN's budget, a significant work-load resulted from CERN's procurement activities related to requirements financed by non-CERN budgets, for example for the major LHC experiments and LHC Grid requirements.

Table 1. Breakdown of main LHC costs, including personnel costs. Figures are in millions of Swiss francs.

	Amounts in MCHF	
	Personnel	Material
LHC machine and experimental areas, Incl. R&D, injectors, tests and pre-operation	1,224	3,756
CERN contribution to detectors, incl. R&D, tests and pre-operation	869	493
CERN contribution to LHC computing	85	83
Total CERN costs	**2,178**	**4,332**

The total LHC procurement activities can be summarized by several key numbers, giving an indication of the challenges faced by the procurement service. The construction of LHC required:

- the issuing of 1,170 price enquiries and invitations to tender (for requirements exceeding 50,000 Swiss francs each);
- the negotiation, drafting and placing of 115,700 purchase orders (for amounts up to 750,000 Swiss francs each) and 1,040 contracts of various types and amounts;
- the commitment of 6,364 different suppliers and contractors (sub-contractors not included).

The total amount can also be broken down into the different main items of the LHC as seen in Figure 3.

As can be understood from the above numbers, the procurement activities related to the LHC project covered everything from orders for a few tens of Swiss francs to contracts exceeding 100 million Swiss francs each, from purchases of a single unit to series manufacturing of hundreds of thousands

3. Total contract expenditure for the LHC project broken down in main components.

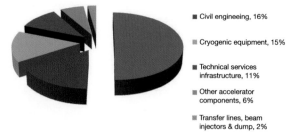

- Civil engineeing, 16%
- Cryogenic equipment, 15%
- Technical services infrastructure, 11%
- Other accelerator components, 6%
- Transfer lines, beam injectors & dump, 2%

of components delivered over several years, from purchases of off-the-shelf items to development and manufacturing of equipment in accordance with specified performance beyond what was considered state-of–the-art at the time the contracts were placed

To describe the strategies applied to the selection of suppliers and contractors, we need first to discuss the implications of the legal status of CERN and to consider the historical approach CERN took with its suppliers, and how these evolved under the constraints of the LHC project.

Legal framework

CERN, an Intergovernmental Organization, was established in July 1953, by the "Convention for the establishment of a European Organization for Nuclear Research". The Convention was initially signed by 12 Member States, but the number of Member States has since increased to 20.[1] It has its seat in Geneva with installations on both sides of the Swiss-French border. CERN has two main sites; the Swiss site (Meyrin) covers approximately 110 hectares and the French site (Prévessin) occupies some 450 hectares. However, even the so called Meyrin (Swiss) site straddles the Swiss-French border, which means that some buildings are in Switzerland and some are in France. This peculiarity sometimes causes some confusion for firms working as contractors on the CERN site, for instance with regard to applicable law. A views of the site straddling the French-Swiss border is found in Figures 3 of Chapter 1.

As an Intergovernmental Organization, CERN is not a legal entity under national law but governed by public international law. The Member States have recognized the international status of CERN (via Host State Agreements with Switzerland and France, and a Protocol on Privileges and Immunities with the other Member States). These agreements ensure that CERN benefits from immunity from national jurisdiction and execution. Thus, legal disputes between CERN and its suppliers and contractors are not submitted to national courts but solved via international arbitration. They also enable CERN to function without interference by individual Member States and guarantee independence from national authorities. For our discussion this means that CERN is thus entitled to establish its own internal rules necessary for its proper functioning, such as the rules under which it purchases equipment and services.

[1] CERN Member States are: Austria, Belgium, Bulgaria, Czech Republic, Denmark, Finland, France, Germany, Greece, Hungary, Italy, The Netherlands, Norway, Poland, Portugal, Slovak Republic, Spain, Sweden, Switzerland and the United Kingdom.

4. The twenty Member States of CERN.

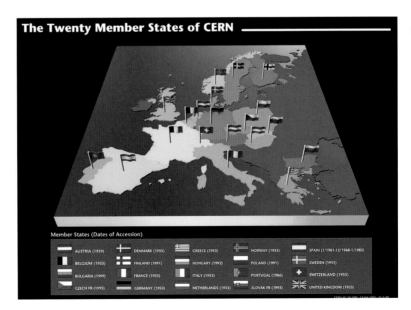

The Twenty Member States of CERN

Member States (Dates of Accession)

AUSTRIA (1959) · DENMARK (1953) · GREECE (1953) · NORWAY (1953) · SPAIN (1/1961-12/1968-1/1983)
BELGIUM (1953) · FINLAND (1991) · HUNGARY (1992) · POLAND (1991) · SWEDEN (1953)
BULGARIA (1999) · FRANCE (1953) · ITALY (1953) · PORTUGAL (1986) · SWITZERLAND (1953)
CZECH FR (1993) · GERMANY (1953) · NETHERLANDS (1953) · SLOVAK FR (1993) · UNITED KINGDOM (1953)

Procurement procedures and national interests

Before the LHC project was approved, the procedures resulting in CERN's award of purchase orders and contracts could be summarized as follows:

- not less than three competitive tenders to be sought for the purchase of plant, equipment, supplies and services;
- invitations to tender to be limited to manufacturers and contractors located within the territories of the Member States;
- contract to be awarded to the firm whose tender satisfactorily complies with the technical, financial and delivery requirements and is the lowest.

However, in the discussions leading up to the approval of the LHC project, the 20 Member States of the Organization, convinced that the distribution of contracts among the Member States was unsatisfactory, made it clear that approval of the project would be subject to a revision of the procurement policy and procedures so as to improve that distribution.

To address this concern, in September 1992, the Organization's Finance Committee, consisting of delegates from all Member States, decided therefore to set up a working group with the objective to "identify ways in which all Member States are given an opportunity of obtaining a fair share of CERN contracts, whilst at the same time ensuring that CERN's procedures do not become too cumbersome and that the Organization obtains best value for money."

In December 1993, the working group submitted a report with a set of proposals that were unanimously approved by the Council after recommendation by the Finance Committee. The most important elements of the new procurement procedures were as follows:

1. The goals of the general procurement policy were defined to be threefold:
- to ensure that bids fulfill all the necessary technical, financial and delivery requirements;
- to keep overall costs for CERN as low as possible;
- to achieve well balanced industrial return coefficients for all the Member States.

2. All CERN contracts were to be divided into two separate classes – supply contracts and industrial service contracts.

3. National interests would be protected by a system of target return coefficients[2], defined for both supply contracts (0.8) and industrial services contracts (0.4).

4. For the purposes of adjudication of supply contracts/industrial services contracts, a Member State was to be considered to be poorly balanced if its supply contract/industrial service contract return coefficient fell below 0.8 and 0.4 respectively, well balanced if it was equal to or greater than that value.

5. For contracts exceeding 200,000 Swiss francs in value, CERN was to apply alignment rules which, under certain well-defined conditions, allowed a bidder offering goods/services originating in poorly balanced Member States to align his price to that of the lowest bidder and thereby be awarded the contract, provided that the bid with the realigned price complied with all the stipulated requirements.

The procedures were implemented and used for all invitations to tender issued from 1994 onwards. The Council further decided that the impact of the revised procedures would be evaluated after three years.

[2] The return coefficient of a Member State for e.g. supply contracts for a given calendar year was defined as the ratio between that Member State's percentage share of all purchases of goods during the preceding three calendar years (excluding energy, fluids, PTT and communication fees and insurance) and that member State's percentage contribution to the budget over the same period.

5. Pieces of ALICE dipole magnet yoke in Dubna, Russia ready for transport to Geneva.

However, a significant part of the contracts placed by CERN were financed, not by the CERN budget, but by funds provided by collaborations of institutes and laboratories participating in the construction of the LHC experiments. As a result of a perceived need for increased financial control of the very large LHC experiments, the Finance Committee requested a financial strategy for these collaborations and their use of the different funds available. The document entitled "Financial guidelines for the LHC experiments" was approved by the Finance Committee in September 1995. Part 2 of that document contained a set of procurement rules and procedures to be applied in four different types of funding arrangements.

- Case A, where no CERN money is involved and the collaboration does not desire CERN's involvement;
- Case B, where no CERN money is involved, but the collaborations do request CERN to be the contract partner on behalf of the collaboration;
- Case C, where CERN contributes 100% of the financial resources for a contract;
- Case D, where contracts are to be funded by a "common fund" established by the collaboration, and to which CERN makes a contribution.

The guidelines stipulated that the rules concerning industrial return and alignment would not apply for contracts funded partly or fully by sources other than CERN (Cases B and D).

In addition to the above procedures, CERN also made special arrangements with a number of non-Member States (e.g. USA, Japan, Russia, India, Pakistan) for the handling of their respective additional contributions, part of which was provided in cash and part as in-kind deliverables.

These measures, and some slight subsequent adjustments resulting from an evaluation of the first three years after their implementation, proved to be fully adequate and resulted in a substantial improvement in the distribution of contracts among Member States. Of great importance, we do not believe that the cost of LHC construction increased because of these measures.

Procurement strategies used for LHC
As can be understood from the previous Section, the procurement activities related to the LHC project required a wide range of strategies. The bulk volume of orders covered straightforward supplies – ordered, delivered and paid. In some areas, however, the tendering and contracting strategies had to be considerably more elaborated. The procurement for the main elements of the LHC can be divided into the following items:

The construction of the LHC

- superconducting magnets with their associated components;
- civil engineering;
- cryogenics.

The tendering and contracting strategies differed among these categories and were based on whether CERN had expertise in the field or not; whether the requirement covered standard products or required extensive development; whether the contracts were based on build-to-print solutions developed by CERN; or whether the contracts were based on functional specifications, etc. It is of course impossible to cover here all the different types of contracts and strategies used for the LHC, but some examples of the different approaches taken will be discussed below for each of these categories.

Superconducting magnets with their associated components

It would be difficult to exaggerate the contractual, technical and logistical challenges in terms of supplying CERN with the necessary magnetic elements. The LHC machine contains some 1,800 superconducting twin-aperture main dipole and quadrupole magnets, along with their ancillary corrector magnets. These magnets are massive and needed to be constructed with absolute precision, and so our approach to their manufacture required creative planning. In addition, as can be seen from Figure 3 above, the total value of the magnetic elements amounts to approximately 50% of the value of the total LHC machine, and the dipole magnets represent some two thirds of this amount.

Creative bargaining: CERN as middleman
The dipole magnets were designed and developed at CERN, although several firms were involved in early prototyping work throughout the development process. The purpose was to get industry involved as early as possible in the manufacturing techniques. In order to gain time and to reduce the risks related to transport, the CERN adopted an unusual policy with respect to the manufacture of the dipole magnets.

As described in Chapter 4, a dipole magnet consists of a *cold mass* (including the shrinking cylinder/He vessel) covered by *superinsulation* and positioned on composite *support posts* inside a *cryostat/ vacuum vessel* (see Fig. 15 of Chap. 1 for an cross-sectional view of the dipole magnet). It was decided at an early stage that the main components and tooling required for the assembly of the dipole cold masses would be purchased by CERN and delivered to the cold mass assemblers. The cold masses, being relatively sturdy equipment, would thereafter be delivered to CERN for final assembly into complete dipole magnets at CERN's premises. Transport of the final dipole magnets (with precisely aligned 30-ton cold masses, resting on thin composite support posts) could therefore be minimized.

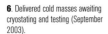

6. Delivered cold masses awaiting cryostating and testing (September 2003).

Although there were contractual complications resulting from CERN's decision to purchase the main components and tooling itself and deliver these to the assemblers, there were many reasons why this was necessary. The main tooling consisted of 15-m-long heavy presses, specially made for curing of the coils and welding of the cold-mass shrinking cylinders. From the date of order, this equipment requires more than a year to manufacture and make operational. If the tooling had been the responsibility of the assemblers, the start-up of cold mass production would have been delayed more than a year, since the assemblers would not have ordered the tooling before having signed contracts for the assembly of the cold masses.

The approach also applies to the superconducting strands and cables needed. The quantities required for the LHC were so huge that all major European manufacturers were involved as well as one US and one Japanese firm. Since the cables were needed in the early stages of the cold mass production, and since the production of the cables required more time than the production of cold masses, the contracts for cables had to be placed before the contracts for the assembly of the cold masses.

For some of the other components (some 20 in total), such as the low-carbon steel laminations in the magnetic yokes and the austenitic stainless steel used for the collars stage of the cold mass production, ans since CERN was primarily concerned about obtaining material with exactly the same properties in all magnets. Since the cold masses were to be assembled by more than one firm, CERN therefore decided to place the contracts for the corresponding material itself, thereby ensuring that all the assemblers used identical material for the magnets and, at the same time, obtaining economies of scale.

Acting as supplier and client

CERN felt that this policy of acting as both supplier and client to external companies was required to keep the project on time and on budget, but this policy did lead to complications. The order for superconducting cable is a good example of the contractual complexities resulting from the chosen strategy. CERN placed contracts with one firm for the supply of niobium-titanium alloy bars and niobium sheets. This material was delivered to four European cable manufacturers with whom CERN had placed contracts for superconducting cable, while the US and Japanese cable contractors procured themselves the same material. The cable was delivered to CERN, tested and delivered to the assemblers. Thus, CERN acted simultaneously both as a supplier and client to the cable manufacturers and the assemblers. This was a delicate operation; in the event of any problem with the dipole magnets, the assemblers might have been tempted to try to blame CERN as a supplier of cable, and the cable manufacturer might try to blame CERN as a supplier of the niobium-titanium alloy bars. Similar contractual set-ups were used for the laminations and collars, where CERN placed contracts for low-carbon steel and stainless steel respectively, and had the material delivered to the fine-blankers with whom CERN had placed contracts for laminations and collars. The laminations and collars were thereafter delivered directly to the assemblers.

7. View of series fabrication of the main quadrupole cold masses.

Although several firms had been involved and interested in the early development phase, at the time of tendering for the dipole cold masses, only three bidders remained committed to the project and had the necessary expertise and experience from prototype manufacturing. However, the production of single units had taken a year or more during the prototyping phase; for the series production, the assemblers would each have to produce three to four units per week! At CERN we feared that this manufacturing time factor would lead to overruns of our budget; bearing in mind the significant risks perceived by the bidders to quote for the assembly of the cold masses, it would not have been inappropriate to add significant risk margins in their tender prices. The fact that probably all three bidders would be needed for the production of the series only aggravated the problem of potentially high prices, since the competition would be reduced.

Creative bargaining part II: assembler contracting
Again, some creative bargaining with the assemblers was called for. Our first idea was to contract each assembler for the production of one (out of eight in total) sector of dipole cold masses. In order to reduce the perceived risks for the assemblers, who had no experience of series manufacturing the dipole cold masses, these first contracts would be based on a target price with incentive fee basis. Thus, each assembler would be rewarded in the case that the final outturn cost was lower than the target price, but he would also contribute to any eventual overrun of the target price. Once the assemblers had gained sufficient experience from the series production to be able to correctly calculate firm prices for the remaining production of dipole cold masses, a new invitation to tender would be issued and new contracts for the remaining five sectors would be placed on a fixed-and-firm-price basis, with two or three of the assemblers.

The call for tenders for the production of the dipole cold masses for one sector (156 cold masses, including two spares) was issued in December 1998. The bidders were asked to give very detailed breakdowns of all cost items in the tender. However, even with this strategy, it was impossible for CERN to agree with the assemblers on a target price compatible with the LHC budget for the dipole cold masses. Furthermore, the strategy would require significant resources for cost auditing of the assemblers, something which was completely new at CERN and for which no procedures or organizational structure existed.

As a consequence, this initial approach failed, and CERN abolished the whole idea of target price and sharing cost overruns or underruns with the assemblers. Instead CERN negotiated firm and fixed prices with all three assemblers for a production of 30 cold masses each. This number, it was estimated, would be sufficient for the assemblers to approach the foot of the learning curve and be in a position to make a reliable estimate of the assembly costs. At the same time, the number was sufficiently low as to reduce the perceived risk for the assemblers, should they have underestimated the costs for assembly.

The negotiations with the assemblers were successful and contracts were subsequently placed in November 1999 with them for the assembly of 30 dipole cold masses each. Towards the end of that production, CERN issued an invitation to tender for the production of the remaining dipole cold masses, requesting the three assemblers to quote firm prices (subject to revision using a price revision formula). The initially quoted prices were again considerably higher than what the budget for the LHC allowed, but after numerous rounds of lengthy negotiations with all three assemblers, during which CERN succeeded in convincing the assemblers that they had overestimated the time required for several assembly operations and with a modification of the payment conditions to better match payments with incurred costs, all three assemblers agreed to prices for the assembly of cold masses that were compatible with the LHC budget. A contract was subsequently placed with each of the three assemblers for a third of the remaining quantities required for the LHC.

8. Cold mass assembly by workers at CERN.

Perils of creative negotiation

The perils of this complex arrangement became apparent when, almost immediately after having paid a significant down payment under the contract, one of the contractors became insolvent. This was completely unexpected and was the result of a clearing arrangement between the assembler concerned and a holding company, which had run out of liquidity. Fortunately, CERN had resisted the requests earlier made by the assembler who had proposed to reduce its price by having the bank guarantee covering the down-payment issued by the holding company rather than by a bank as requested by CERN, in which case the down payment would not have been recoverable. CERN could therefore recuperate the funds lost in the insolvency and negotiated a new contract with the restructured assembler.

In the end, the strategy of CERN proved to be sound. The assembly of the dipole cold masses, with their tooling, materials and components provided by CERN, was considered to be the most challenging aspect of the LHC project. Our contractual strategy worked on all levels – technical, financial, contractual and logistical – and neither delayed the LHC project, nor resulted in significant cost overruns. In this story, we see lessons for the future in terms of the negotiation of equipment procurement for large international scientific collaborations.

Civil engineering

The value of all civil-engineering activities for the LHC project was approximately 500 million Swiss francs. Time constraints, supplier risk, and lack of internal CERN resources called for a novel approach to the contracting for civil engineering, one in which contract allotment and inspiration from a standard contractual framework would both play important roles.

Due to significant reductions of CERN's civil engineering group, it was from the start decided that CERN would not have the resources required for the design and supervision of the works. A conventional division of tasks into three parties: client (CERN), consultant and performing (or construction) contractors would therefore have to be adopted. Under this scheme, the consultants, selected among international design engineering consortia, were entrusted with comprehensive assignments (design studies, preparation of calls for tenders for the work, and supervision of work execution).

For their part, the consortia performing the construction works took over the performance of the work, using price lists (bill of quantities) and drawings produced by the consultants. For certain work packages, however, it was decided to give them additional responsibility for preparing the detailed working drawings.

Organization of engineering work into packages
One of the earliest decisions taken, around 1994, during the preliminary reflections on the LHC project, was to divide the civil engineering project into *packages*, with the aim of:
- limiting the risks in the event of difficulties with any single consultant or contractor; and
- creating work packages of a sufficient size to interest consultants and contractors of international standing, with the capacity and references commensurate with such an ambitious project.

For the previous LEP project, the surface and underground work had been kept separate and contracted out to specialized consortia. Thus, the "surface" consortia took over work sites in stages, as their counterparts responsible for the underground structures completed their work. This process was plagued by friction regarding completion delays, access sharing, and pollution and waste removal.

CERN drew the lesson from that experience and decided this time to divide the new project into three large packages along strictly geographic lines:
Package 1: the buildings, related structures and underground structures in the area of the detector ATLAS (Point 1);
Package 2: the buildings, related structures and underground structures in the area of the detector CMS (Point 5);
Package 3: the remaining surface and underground work around the ring, including the caverns and transfer tunnels TI2 and TI8 (seen in Figure 10) and the beam dumps.

9. CMS cavern under construction as part of Package 2.

10. Construction on the LHC transfer tunnel.

In February 1995, CERN issued the preliminary inquiries in view of qualifying bidders for the design. A jury of four experts (three non-CERN experts and one CERN expert) conducted the selection process, which resulted in a list of 17 consortia that were to be invited to take part in a call for tenders conducted on the basis of the "double envelope" procedure. In accordance with this procedure, the tenders are submitted in two separately sealed envelopes – one for the technical documentation and another for the commercial aspects. The idea is to first evaluate the technical proposals without knowing the prices and thereafter only open the price information submitted by the bidders having being considered technically qualified.

The technical envelope comprised administrative information, the work programs, proposed methods, equipment characteristics, the manpower proposed, a list of sub-contractors together with evidence that they fulfilled the qualification criteria. The commercial envelope consisted essentially of overall prices, financial conditions, a detailed price breakdown, and payment schedules. Once tenders were received in November 1995, the jury of experts met again to examine the contents of the technical envelopes and selected the consortia whose tenders fully met the technical specifications of the call for tenders. Finally, the financial envelopes from those consortia were opened in January 1996, and the contracts awarded to the lowest bidders in April 1996, and the consultants were able to start design work immediately after.

Shortly after design work commenced in the spring of 1996, the Swiss Confederation decided to finance the work for injection tunnel TI8, as a special host-state contribution to the LHC project. As the work had to be performed by Swiss companies, Package 3 was at this point divided into two sub-packages: 3b for tunnel TI8, and 3a for all other Package 3 work. Another additional package was later defined for a contract covering work for CNGS (CERN Neutrinos to Gran Sasso, a project that aims at measuring the oscillations of neutrinos).

The design consultants prepared the technical parts of the invitations to tender (specifications, drawings and price lists), and the selection procedure for the construction contractors was able to get underway in August 1996. This was organized in the same manner as the selection of the design consultants. Thus, the tendering for the three packages was to be based on a double-envelope principle, whereby the technical part of the bids was to be evaluated first, by a panel of internal and external experts.

The selection jury, consisting of four experts in large-scale underground construction, was convened twice, and only the financial envelopes of those firms whose technical bids had been declared acceptable by the jury were thereafter opened and contracts could then be awarded in accordance with CERN's procurement procedures. The three winning consortia (for Packages 1, 2 and 3a) were awarded contracts in February 1998. Package 3b was subject to a separate selection procedure, which was organized and conducted with the participation of the Geneva cantonal authorities.

Contracting the suppliers
CERN's standard general conditions were not suitable for civil engineering contracts of this magnitude, and so it was early decided to use the internation-

The construction of the LHC

ally recognized *Conditions of Contract for Works of Civil Engineering Construction*, the so-called "Red Book", edition 1987, issued by the International Federation of Consulting Engineers (FIDIC) as a basis for the LHC civil engineering construction contracts. However, the Red Book was adapted for the purpose of the LHC contracts, in particular so as to reflect CERN's status as an Intergovernmental Organization and its internal decision-making procedures. Thus, for instance, a number of approval duties (in particular concerning final approvals of documents, payments and decisions in relation to claims and extensions of time) were transferred from the consultant to CERN. In addition, it was decided to refer disputes, if any, to an adjudication panel of independent experts with a view to obtaining settlements within a few weeks and subsequently, if a party was dissatisfied, to arbitration. However, the parties had to accept the adjudication panel's decision and had to continue executing their contractual obligations until termination of the contracts. Only then could a party ask for arbitration. It is interesting to notice here that later editions of the Red Book include provisions regarding such an adjudication panel, similar to those implemented by CERN.

The contracts clearly defined the allocation of risk between CERN and the contractors according to the standards of the Red Book. CERN's share of the major risks included:

- ground or other physical conditions being different from those described in the tender documents;
- adverse climatic conditions;
- force majeure;
- legislation changes in the country where the construction takes place;
- changes in work quantities from those described in the tender documents;
- changes in planning by CERN.

Adjusting contractual terms in real time

Perhaps not too surprisingly for a project of this magnitude, during the actual construction, a number of events occurred that led to the necessity to amend construction contracts. The major events are described in some detail in the

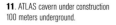

11. ATLAS cavern under construction 100 meters underground.

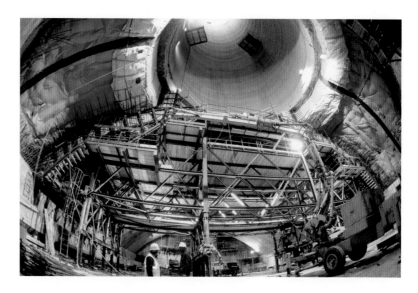

second part of this Chapter; we summarize these briefly below from the managerial viewpoint:

Package 1: In this package, related to the construction of the ATLAS cavern, the main events were (a) a significant underestimation of the quantities used in the invitation to tender; (b) the discovery that the design of the vault of the cavern using reinforced concrete beams was technically unacceptable and that a completely new design and construction method was necessary; (c) changes in Swiss legislation; and (d) some additional requirements by CERN. CERN and the Contractor could settle the issues without involving the adjudication panel.

Package 2: In this package, related to works for the CMS shafts and cavern, the main events were (a) unforeseeable ground conditions encountered during the construction delayed the ground-freezing operations considerably; (b) design modifications were required after the geological conditions in the underground works were found to be worse than expected from the geotechnical investigations and which resulted in significant changes to the construction of the works; and (c) changes in French legislation. These events were important enough to lead to the submission of disputes to the adjudication panel. However, a final satisfactory settlement was reached at the end of the contract.

Package 3a: As regards the various tunnel works, a delay in the notification of the commencement date by CERN meant that the planning on which the contract had been based was no longer valid, which implied significant changes in required equipment and manpower. In addition, unforeseeable ground conditions slowed progress down and, as a consequence of the ongoing development of the LHC machine, CERN made several important changes to the scope of work. The result was that the contract with its bill of quantities, which was based on a completely different scenario, became almost inoperable. Instead of following an agreed planning and schedule, the work progressed on daily ad-hoc instructions by the consultant to the contractor, which made it impossible to evaluate costs and plan the work correctly and the relations with the contractor started to deteriorate.

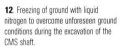

12. Freezing of ground with liquid nitrogen to overcome unforeseen ground conditions during the excavation of the CMS shaft.

In order to solve the problem, CERN and the contractor for Package 3a agreed to change completely the contractual basis for the works. Based on the rates and prices given in the contractor's original tender and the above revised requirements of CERN in terms of planning and scope of work for the underground civil-engineering work, a first maximum cost-to-completion estimate was established and agreed upon. In order to create incentives for the contractor to find ways to reduce the estimated cost-to-completion, CERN and the contractor established a "Supplemental Agreement" which took into account the revised requirements. In the "Supplemental Agreement" the method of calculating the remuneration was modified such that the contractor would be paid for actual costs incurred, rather than in accordance with the initial contractual bill of quantities. In addition, the revised conditions stipulated that the contractor would be rewarded in the case that the final outturn cost was lower than the cost-to-completion estimate and that the he would contribute to any eventual overrun of the cost-to-completion estimate. With the "Supplemental Agreement" as a basis, it was thereafter possible to collaborate actively with the contractor with a view to reducing the cost to completion and proceed with the works as fast as possible.

In all cases (Packages 1, 2,3a and 3b), despite the technical difficulties encountered by the contractors and the contractual challenges, all disputes were successfully solved, either between the parties directly or, in few cases, with the help of the adjudication panel. In no case did any party request arbitration at the end of the construction. Thus the civil engineering phase of the project was completed to the satisfaction of all the parties, both in terms of deadlines and costs.

Cryogenics

The superconducting magnets referred to above operate in superfluid helium at a temperature of 1.9 K and require cryogenic installations on an unprecedented scale, to be constructed around the whole circumference of the LHC machine. Refrigeration at cryogenic temperatures is produced by eight large refrigerators, each of which supplies an LHC sector. To distribute the cryogenic

13. A view of the 27-km cryogenic distribution line before the installation of the dipole magnets.

refrigeration power, helium is transported over long distances and distributed to the local magnet cooling loops via a 27-km-long cryogenic distribution line (referred to as the QRL). A detailed discussion of the cryogenic challenges of the LHC is found in Chapter 4.3.

Since there were several firms in the CERN Member States competent in design, production and installation of cryogenic transfer lines, CERN felt that it was not necessary to design the QRL itself. However, it was clear that, asking bidders to quote competitive prices for the complete QRL before any design or prototyping work would result in prices that would include significant risk margins, since the risks perceived by the bidders for such a big installation and with such stringent technical requirements would be too important. Again, CERN needed to build a strategy for supplier negotiation that would reduce the perceived risk and keep the production on time and on cost. These discussions were based on the following two-stage approach:

Stage 1 (Qualifying pre-series test cell): Based on a functional specification, bidders were invited to submit bids for the QRL design, manufacturing, installation, and commissioning of a 110-m-long pre-series test cell, to be located at CERN, and also to quote net ceiling prices for lots composed of one to eight 3.3-km-long sectors of the final QRL. The adjudication of the pre-series contracts was based on the sum of the price of the pre-series test cell and that of half the price for all eight sectors (with the idea of working with at least two suppliers). The contract(s) for the series QRL would be subject to a second round of bidding among the successful bidders whose pre-series test cell had been technically qualified by CERN.

Stage 2 (Series QRL): After delivery and successful testing of the pre-series test cells, the suppliers were invited to re-tender for the series QRL, with the stipulation that tenders must not exceed the net ceiling prices quoted for the series QRL in the first tendering round (Stage 1).

From three to a single contractor

As a result of the tendering exercise in Stage 1, CERN consequently placed contracts with three contractors, each for the design, manufacturing, installation at CERN and commissioning of a 110-m-long pre-series test cell. All three contractors produced a test cell meeting CERN's requirements and were invited to submit revised bids for the supply and installation of the QRL. Initially, it was our intention to split the total requirement between two suppliers, with a part of the eight sectors from each, thus following a dual-sourcing policy as a provision against supply difficulties of one of the contractors. Bidders were therefore informed that the adjudication would be based on the average price quoted for four sectors of the QRL (i.e., for 50% of the complete QRL). However, the bidders were invited to quote separately on the price per sector for the construction of a larger number of sectors.

To our surprise, one of the bidders, who successfully passed the pre-series test, quoted a very advantageous price for the entire set. CERN thus decided not to split the order, but to place a contract for the complete QRL with this firm only. The contract was signed in June 2002. The contract was based on fixed prices per sector of QRL (the eight sectors were not identical) and included a price revision formula.

Work began quickly, but by the spring of 2003 a number of problems came to light. These included the delivery of non-conforming components by a sub-contractor; a change in sub-contractor for the installation of the QRL in the LHC tunnel, who subsequently became insolvent; and corrosion caused by

14. Disassembled sections of the cryogenic transfer line ready for repair.

the flux employed for brazing operations. This situation created serious concern among the LHC management because it delayed work enough to have a knock-on effect on the rest of the LHC project schedule. For example, the magnet installation in the tunnel could not start before the QRL had been installed and tested in the concerned sectors.

CERN rescues its supplier
CERN was forced to take a decisive – but rather unusual – course of action in order to gain lost time in the production and installation of the QRL. Firstly, CERN agreed to pay the contractor certain costs related to the speeding of component production and to additional tooling and additional supervisory manpower required for accelerated installation. Secondly, CERN agreed to assume a number of tasks that had initially been assigned to the contractor. For example, a contractual amendment was signed, by which CERN agreed to take over responsibility for installation of sector 7-8 (the first QRL sector installed by the contractor), thus "freeing" the contractor to focus more on the rest of its work. Much of the sector had already been installed by the contractor, but it was found to be defective (and behind schedule). Taking over sector 7-8, meant that CERN had to remove the installed lengths, repair them (as well as the faulty elements in stock yet to be installed), and (re)install them. This would allow the contractor to concentrate on the production of elements and installation of the other seven sectors, without having to interrupt his production lines for the repair of faulty components.

As a result of these exceptional efforts, the installation accelerated considerably and, despite the short installation time of the final sectors, the quality of these sectors was high. The implemented changes resulted in lengthy discussions and negotiations between the parties as regards extra costs incurred by both parties but an overall settlement was reached at the end of the contract.

Striking the balance with pressured suppliers
It is important to place the individual events and actions described above in the context of the overall project. CERN had to balance many conflicting interests whilst dealing with the problems associated with the QRL. Given CERN's international role and reputation, there was undoubtedly intense pressure for the LHC project to come in on time. The target date was being jeopardized by the problems with the QRL, but CERN had no alternative contractor to turn to. In the circumstances, CERN was unable to apply too much pressure on the contractor, as the situation would have been irreparably escalated if the contractor were to decide to "down tools" or bring arbitral proceedings before completing his work. In fact, it CERN's chosen approach to collaborate with the contractor turned out more efficient to achieve successful completion of the works.

The lessons learned

In this first part of Chapter 3, we looked in some detail at the most costly contracts of the LHC construction. With the few examples detailed in this chapter, we hoped to convey the innovative style of interacting with suppliers and contractors that allowed the LHC to be built only very slightly over budget. In the context of a large international scientific collaboration, it is critical to make sure that governmental support, as presented at the start of the project, is not undermined because of cost overruns continuous delays.

The lessons learned from our experience can be summarized in the following points:

• Different contracts require different tendering and contracting strategies. In the examples we saw three distinct approaches; (a) we used international recognized contract standards for the civil engineering; (b) the dipole cold masses were developed at CERN and the contracts based on build-to-print drawings, and CERN even supplied the main tooling and all major components for the assembly; and (c) the QRL contract was based on a functional performance specification where the design and development was performed by the supplier.

• Flexibility and innovation in the procurement process is of utmost importance. If a strategy doesn't work as planned because of unforeseeable conditions, you have to be prepared to change strategy. In the case of LHC we changed the tendering strategy for the dipole cold masses in the middle of the tendering process and we changed the civil engineering contract approach after having started the work. These changes, although difficult and complicated at the time, turned out to have been instrumental in getting the budgets under control.

• Carefully evaluate the benefits versus costs related to dual sourcing.

• Finally, CERN's procurement strategies were all based on the fact the technical expertise and competence existed in-house in all relevant fields; cryogenics, superconductivity, ultra-high-vacuum application, electronics, computing, and civil engineering.

15. Transport of the ATLAS barrel toroid to CERN. (Photo: Peter Ginter)

3.2

THE CONSTRUCTION OF THE LHC

Civil Engineering Highlights
Jean-Luc Baldy,
Luz Anastasia Lopez- Hernandez
and John Osborne

From the point of view of the civil engineer, the construction of the LHC represented a series of unique challenges, some of which were foreseen while others, as is the case with all projects of this magnitude, could not have been anticipated. The starting point was the existing tunnel built for the Large Electron Positron (LEP) machine, constructed between 1984 and 1989. Great care has been taken to re-use, as much as possible, the existing civil engineering infrastructure that was created for the LEP machine and in fact all LEP buildings, some with modifications, are now being re-used for the LHC. In this chapter, we will present an overview of the construction works, after which we will discuss the general needs and modifications of the underground and above-ground structures. Some of the more challenging aspects of the construction are highlighted, with the hope that this will convey to the reader the scale and complexity of the task. We finish the chapter with a few comments on safety and environmental aspects.

The challenge at hand

The focus of the civil engineering work was of course to supply the scientists with the infrastructure required for the hadron beam and detectors. As for the LEP, the equipment has to be housed in an underground tunnel because of the sub-millimeter positional stability that the machine requires. Indeed, the civil engineering was not a minor component of the LHC project:

- the basic tunnel configuration was already established, but more room was clearly needed than supplied by the existing LEP infrastructure;
- caverns and tunnels had to be excavated taking the geology of the site, including the presence of hydrocarbons and shifting aquifers, into consideration;
- huge amounts of sensitive equipment would have to be lowered into position; existing structures would have to be re-equipped;
- the protection of the environment and of the safety of workers and local inhabitants would remain a priority;

In summary, the scientists required a main beam tunnel, with access shafts from the surface to the underground areas, together with various underground caverns and other ancillary structures for housing equipment that cannot be located on the surface. On the surface, buildings were required for housing compressors, ventilation equipment, electrical equipment, access control, and control electronics. New underground equipment to be added in-

The construction of the LHC

cluded the new injection lines and the beam dumps. In addition several small caverns were required to house equipment at various locations around the main beam tunnel.

Of the four LHC experimental areas, two have been constructed on almost "green field" sites in that there was very little existing infrastructure. As such, two large experimental zones for ATLAS and CMS had to be constructed at Points 1 and 5 respectively, as shown in Figure 16, where the concentration of red underscores the extent of the new construction. These two new experimental zones are similar in that they both consist of two new large caverns, one for the detector and one for the services, together with various galleries, tunnels and chambers for housing equipment and providing access routes. For the two smaller experiments (ALICE and LHCb) the existing infrastructure required only minor modifications to accommodate the new detectors.

Timeframe and planning of the civil engineering works

Most of the civil engineering works for the LHC Project lasted 4.5 to 5 years, except for Point 5 (CMS experimental area) where it took approximately 6.5 years to complete. The schedule is summarized in Figure 17. It is of interest to note that the final cost of the civil engineering project has been approximately 498 million Swiss francs, of which 50 million francs were for the consultants, experts and architects, and 448 million francs for the construction work itself.

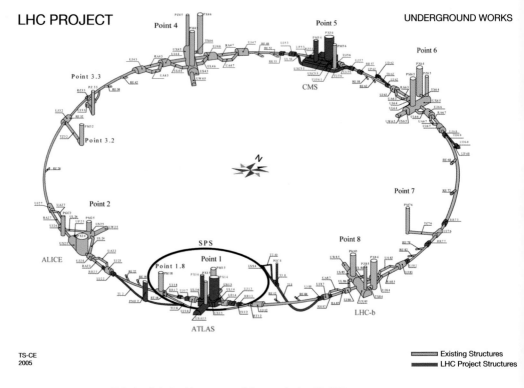

LHC PROJECT

UNDERGROUND WORKS

TS-CE
2005

■ Existing Structures
■ LHC Project Structures

16. A schematic drawing of the new structures added to the existing LEP infrastructure for the construction of the LHC. The new elements are shown in red and are concentrated at the Points 1 and 5, the construction sites of the ATLAS and CMS experiments. Other notable additions include the beam transfer tunnels from the SPS (near Point 1) and the beam dumps (around Point 6).

17. An early schedule highlighting the extent of the civil engineering works that were accomplished while the LEP was still in operation.

LHC CIVIL ENGINEERING	1998	1999	2000	2001	2002	2003	2004	2005
Point 1 - **Atlas**								
Point 1.8 - Prevessin (Surface buildings)								
Point 2 - **Alice**								
Point 4 - Echenevex (Surface buildings)								
Point 5 - **CMS**								
Point 6 - Versonnex (Beam dumps & Surface buildings)				LEP DISMANTLING				
Point 7 - Ornex (RZ tunnel enlargements)								
Point 8 - **LHC-b**								
TI 2 - Injection Tunnel								
TI 8 - Injection Tunnel								

Until November 2000 when CERN decided to stop and dismantle the LEP accelerator, the civil engineering work had to be carried out with the least possible disturbance to the operation of this accelerator (for instance all blasting was prohibited for underground excavations, precautions were taken to minimize the infiltration of dust in the LEP areas and movements of the existing LEP structures had to be minimized). After November 2000, these precautions could be lightened but the constraints from the general LHC programme meant that a lot of work had to be carried out in parallel.

One early complication was the French "Declaration of Public Utility" (DUP) procedure, which must be obtained for all project work sites in France, and for which an Impact Study had been submitted in advance. With the full agreement of the Swiss authorities, the same study covered the extent of the project in Swiss territory. While the approval of the Swiss authorities was obtained at the start of 1998, the DUP was not approved until 30 July 1998. Construction on French territory therefore started with delays of six to eight months. For the injection tunnel, the delay seriously disrupted the work schedule, which had been established around the planned stoppage dates for CERN's LEP and SPS machines, which had not changed.

After extensive negotiations, CERN was able to reach an agreement on delivery dates for the different structures of the LHC project that satisfied the LHC Project Management Unit, took account of the constraints of the other participants, and complied with the contract provisions:

- The main structure for the ATLAS cavern was to be delivered to the installers in mid-April 2003, and the last building in this area in mid-June of that year.
- The cavern for the future CMS detector, was to be handed over to the physicists in early 2005, while the final buildings, including the one housing the experiment's control room, were to be delivered in mid-May 2005.
- For the TI2 injection tunnel and associated structures, in particular the two large magnet transit buildings above the access shaft, were to be handed over to CERN at the end of January 2003. The other major structures in this package, those for the beam dumps, were to be finished at the end of April 2004.
- All the structures relating to tunnel TI8 were to be completed in mid-October 2002.

In spite of several unanticipated events, we were able to meet these ambitious deadlines – this chapter tells the story of how a certain number of these challenges were met.

Underground structures

Before describing the LHC structures themselves, it should be mentioned that the LEP tunnel, 26.7 kilometres in circumference, was built at depths of between 45 and 170 meters, and lies in a plane that is inclined at 1.4 degrees

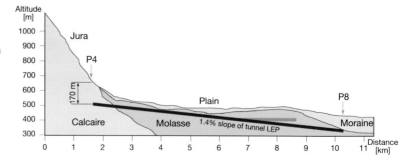

18. Schematic section of the regional geology, showing the 1.4% tilt of the LHC tunnel ring as it lies within the molasse (a sort of sandstone) under the Swiss plain. Points 4 and 8 are indicated with arrows.

from the horizontal; this was done so that the tunnel would lie almost entirely in the Leman basin *molasse*.

This was dictated by the need to ensure that the machines installed in the tunnels, with components aligned to one-tenth of a millimeter, remained perfectly stable. The molasse is a veritable layer-cake of alternating marl and sandstone, the individual layers having a thickness in the order of 10 centimeters in places. The hardness of the molasses also varies considerably, further complicating the work.

Details of the new work for the LHC
Although the LHC re-used most of the underground structures built for its predecessor, a number of major underground construction projects were required:

- Entirely new experimental areas, dwarfing those for the LEP experiments, had to be built for ATLAS and CMS at Points 1 and 5 of the ring, respectively. This necessitated the excavation of several very large caverns and access shafts.
- To transfer the LHC's proton beams, consisting of particles that weigh 2000 times more than the electrons and positrons of the LEP, new transfer tunnels were also required. This is indicated in Figure 16 as the red tunnels leading from the SPS near Point 1.
- Once their quality has begun to degrade (after roughly 24 hours of operation), the beams will have to be stopped with massive steel-graphite beam dumps housed in specially constructed caverns. These are seen as the red tunnels branching off the ring at Point 6 of Figure 16.

19. Showing the position of three test borings near the CMS cavern. Note that the caverns and tunnel are deeply imbedded into the molasse, and that the overall depth is about 100 meters. The variability of the local conditions is also highlighted.

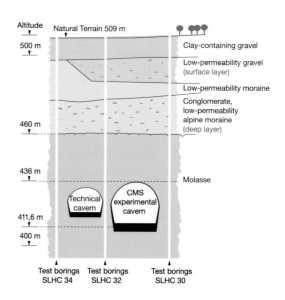

Despite the considerable amount of information already available on the subsurface geology, a legacy of LEP construction, extensive geotechnical investigations were conducted for this project. The investigations involved making 34 test borings, representing a total of over 30,000 meters of borings. Numerous field and laboratory tests and measurements were conducted, involving some ten firms and institutes.

The basic parameters sought were those relating to
- the location of the top of the molasse;
- the groundwater flow rate in the moraines, measured using tracers;
- field stresses;
- rock strength, swelling and abrasiveness.

The position of three test borings around the site of the CMS cavern is shown in Figure 19, where it is seen that the underground construction is imbedded into the local molasse.

In some cases this testing proved invaluable, as in the case of the design studies for the injection tunnel TI2, where it led to modifications to the longitudinal profile of the tunnel to take into account a depression in the moraine/molasse interface under the French commune of St. Genis-Pouilly. The final configuration of the tunnel is drawn in Figure 61 of Chapter 4, where it can be seen how the trajectory of the tunnel was modified to maintain its position entirely in the molasse. In other cases, the studies confirmed the presence of natural hydrocarbons (heavy petroleum) in some areas of the project (tunnel TI8, ATLAS detector cavern). This made it possible to take them into account in contractual arrangements and to make the necessary preparations, both technical and administrative, for dealing with them.

20. A view of the use of hydraulic rock-breakers during the construction of the ATLAS cavern.

21. The head of the tunnel boring machine that was lowered for the construction of tunnel TI8.

22. The application of "shotcrete" to reinforce a subterranean cavern.

23. A view of the welded lattice reinforcement used in the CMS cavern.

Working in the molasse of the Franco-Swiss plain
CERN had from the outset ruled out the use of explosives for all the project excavation work (except as a last resort). There were two main reasons for this decision, one of them external (the need to protect the inhabited areas around the CERN sites) and the other internal (the need to avoid disruption to CERN facilities, in particular LEP accelerator operations, which continued until November 2000). The contractors performing excavations in the molasse accordingly relied on a range of conventional equipment:
- hydraulic rock-breakers for most of the caverns (Fig. 20);
- road headers for tunnels and cavern finishing work;
- tunnel boring machine for tunnel TI8 (Fig. 21).
- temporary support for in-molasse construction, relying on the "New Austrian Tunnelling method" throughout, with some variations between packages and structures;
    - application of sprayed concrete (shotcrete) varying in overall thickness from 75 to 500 millimeters, with or without fiber reinforcement (Fig. 22);
    - welded lattice reinforcement in some cases, in some cases imbedded into the shotcrete (Fig. 23);
    - rock bolts 25 to 40 millimeters in diameter and 1.50 to 12 metres in length;
    - in certain cases, Swellex expansion-type rock bolts were used.

The ATLAS, CMS, and certain tunnel sites represented specific challenges; we present these in separate sections below. The ATLAS situation high-

lights the resourcefulness of the engineering team to design their construction site under the strictest specifications imaginable; the CMS job shows how an unforeseen force of nature had to be overcome by "on-the-fly" engineering.

ATLAS construction site

In the case of ATLAS, two caverns, arranged at a right angle, needed to be constructed. The finished dimensions of the main cavern are impressive, with a length of 53 meters, a width of 30 meters, and a roof height of 35 meters. The service cavern is smaller, with a diameter of 20 meters and a length of 65 meters. The two caverns are connected by five L-shaped tunnels, measuring 2.20 to 3.80 meters in diameter. Two smaller caverns have been excavated to house the electrical equipment for the LHC machine; they are situated to either side of the experimental area on the existing tunnel.

This area is located at Point 1 of the accelerator, opposite the main entrance to CERN's Meyrin site (Switzerland). As the top of the molasse extends to within six or seven metres below the surface at this point, no particular problems were encountered in excavating the vertical access shafts.

However, the excavation and concreting work of the ATLAS large cavern proved to be a real challenge. The users of the LEP accelerator, whose tunnel runs through the middle of the cavern, insisted that the excavation of the upper part of the cavern must under no circumstances result in an upward displacement of the top of the beam tunnel by more than 30 millimeters. To avoid lengthening the critical path for the work package, it was decided that initially only the top 10 meters beneath the crown of the cavern would be excavated, based on the results of a simulation conducted using a three-dimensional model. In this way it was possible to finish the concrete work on the cavern roof, while LEP lived out the final weeks of its operational life, largely unperturbed, 17 meters below. The next problem was ensuring the stability of the roof, which also included the future beams for the overhead travelling crane and the upper portions of the two end-walls.

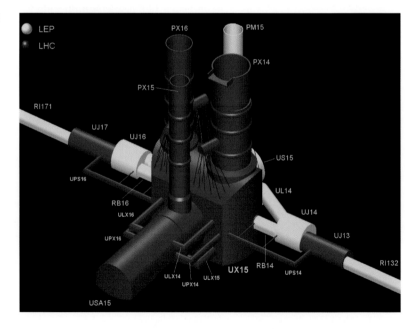

24. The axonometry of the underground ATLAS experimental area, showing the position of the supporting cables.

25. Images of the use of stranded cabling as suspension for the roof of the ATLAS cavern; (a) cable as extending through the roof ceiling; and (b) the secured upper ends of the cables within a gallery.

26. The ATLAS cavern suspended concrete vault.

After examining several different options, which was further complicated by the position of the ATLAS cavern between an existing cavern and its own service cavern, the team of engineers came up with an original solution. It involved suspending the roughly 8,000 tons of concrete using 38 thirteen-stranded cables running from different points of the roof to four small galleries located 20 meters above, which would be constructed from the two access shafts. The cables were positioned before the concrete was poured. Finally, the cables were tensioned to 220 tons from the four galleries holding the active anchorage points. Some photographs of this engineering feat are shown in Figure 25. A view of the suspended ATLAS vault is shown in Figure 26.

The entire operation went off exactly as planned. In March 2002, some nine months after the work was done, measurements showed that the roof had moved – amazingly – by no more than about one millimeter, confirming the soundness of the scheme. After the concrete for the cavern's five-meter thick base slab and for its two-meter thick wall has been poured, grout injection was used to fix the crane rail beams, at which point the tension was taken off the cables and anchors.

CMS construction site

For the CMS detector, the two principal caverns were placed in parallel to keep connections as short as possible. The finished experimental cavern measures 26 by 53 meters, while the service cavern is 18 by 85 meters. As the distance between the two caverns is fixed at approximately 7 meters, the section of molasse between them had to be replaced by a concrete pillar 50 meters long and 28 meters high. This work was done first, after the two shafts had been bored and before the two large caverns were excavated.

This area is situated near the French village of Cessy, in the Pays de Gex. Unlike the ATLAS area, the moraine-molasse interface is situated at a depth of more than 50 meters at this location, in other words only 18 meters above the cavern vaults, on average (Fig. 19). There are also two aquifers in the moraines

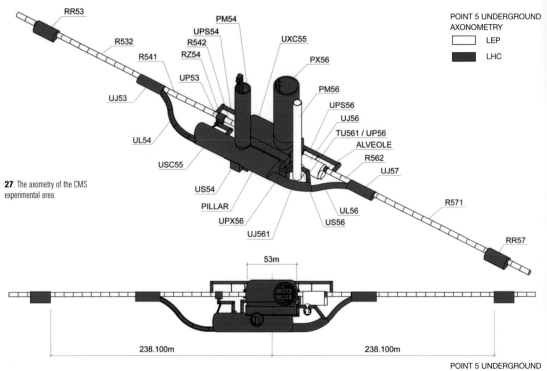

POINT 5 UNDERGROUND
AXONOMETRY

▢ LEP

■ LHC

27. The axometry of the CMS experimental area.

POINT 5 UNDERGROUND
LAYOUT

at different levels, further complicating the construction of the two access shafts, as will be seen below. As for the ATLAS area, ancillary tunnels and caverns need to be built, including a complete bypass via the service cavern.

Different solutions were found for the various aspects of the construction site. In the case of the vulnerable shaft-cavern intersection, as at Point 5, anchored rings were built up with shotcrete and sprayed onto the molasse with 15-m-long pre-stressed ties. Also, tunnel-bore machine (TBM) operations for tunnel TI8 encountered numerous obstacles. For example, rock fall (up to twenty cubic meters), both in the roof and the sides of the tunnel, stopped the machine's advance on several occasions. In general, such falls occurred where molasse strata had been weakened or degraded, in some cases as a result of inadequate overhead shoring. In some sections, the presence of hydrocarbons formed a sludge with the molasse debris at the machine head and slowed progress of the bore.

However, the most challenging stretch of the CMS construction site involved the excavation of the two access shafts at Point 5, where we needed to pass through 50 to 55 meters of groundwater-bearing moraine to reach the top of the molasse. Two options, deep diaphragm walls and ground freezing, were studied over a period of several months. A conventional ground freezing procedure was employed, with a primary cooling circuit using ammonia and a secondary circuit using brine at $-23\,°C$, circulating in vertical tubes in pre-drilled holes at 1.5-meter intervals.

There were two moraine aquifers, at an approximate depth of 15 and 40 meters respectively, with local groundwater flow rates of up to 20 meters per day, which initially overwhelmed the engineers efforts.

After freezing commenced, problems arose in both shafts with "windows" opening up in the frozen soil wall. The subcontractor handling the ground

28. Views of the CMS shaft construction,
(a) schematic drawing showing the
linking up cylinders of ice to construct a
temporary wall; and (b) image of the use
of liquid nitrogen to freeze the soil.

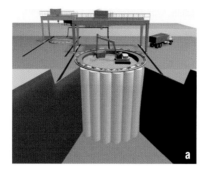

freezing work tried several ways of remedying the problem: additional injections, more brine tubes, bentonite cement injections to reduce the flow rates, and using liquid nitrogen in some of the existing tubes. In both cases, the firm finally managed to seal the frozen wall, but at the cost of a roughly five-month delay on the critical path, and cost overruns that resulted in a large claim against CERN.

A schematic of the technique is shown in Figure 28(a); and a photograph of the site can be seen in Figure 28(b).

Transfer Tunnels and Beam Dump

Additional infrastructure was required to allow the injection and removal of the circulating protons. The protons for the LHC are provided by the Super Proton Synchrotron (SPS) which is located near the ATLAS experiment at Point 1, where two transfer tunnels were built. Also, at the end of a run or in the event of a malfunction, the beam must be "dumped" (as described in Chap. 4.4); for this two tunnels were designed as well.

The two transfer tunnels for the proton beams between the SPS and LHC accelerators have the same finished diameter of 3 meters, and roughly the same length of about 2.5 kilometers. The first, named TI2, connects Point 2 of the LHC to the 8-km-circumference SPS accelerator. It is of interest to note that the access shaft is the only one on this project to have an elliptical cross-section (12 by 8 meters). It was used to lower the machine's 1,330 16-m-long dipole magnets in horizontal position. The tunnel's longitudinal profile had to be modified during the design process, to allow for a dip in the upper limit of the moraine and an aquifer underneath the French commune of St. Genis-Pouilly (see Fig. 61 of Chap. 4). The second tunnel connects point 4 of the SPS accelerator to Point 8 of the future LHC; the tunnel already had an access shaft situated at the SPS end. This tunnel has a relatively constant slope of 3.8%, and lies 50 to 100 meters below the surface. A photograph of the start of the tunnel work is shown in Figure 10.

The structures needed for the beam dumps are situated on either side of Point 6, near the French village of Versonnex. For the beams to be ejected tangentially and to obtain the distance required for their dispersal, it is necessary to construct an extraction cavern (8 by 20 m), a tunnel (3-m diameter and 330-m in length), and an end cavern (9 by 25 m), joined to the existing tunnel by a short gallery. Environmental and economic considerations led to the decision to carry out this work from the existing shafts at Point 6, rather than constructing a new access from the surface. Some more details about these efforts are provided in Chapter 4.4.

Subterranean structures: final notes

Underground construction represented the most challenging aspects of the LHC civil engineering. Thanks to careful planning and some improvisational engineering, we managed to keep the construction projects more or less on time and on budget. For reasons of economy, the number of other underground structures to be realized on the existing ring has been reduced to the absolute minimum. Thus, apart from a few new foundations for the ALICE detector in its cavern at Point 2, some concrete cutting operations and reinforcement of the end wall of the Point 8 cavern (housing the LHCb detector) to deal with unexpectedly high pressure from the molasse, the additional work not mentioned in this section is limited to the excavation of two caverns at Point 7.

Surface Buildings

Surface construction is far less of an engineering challenge; we briefly summarize the modifications that were carried out to the existing structures (inherited from the LEP experiment). A total 30 new buildings were erected in the frame of the LHC Project, representing a total gross area of 28,000 m²:

- construction of a clean room in building SXL2 and a counting room in building SX2;
- at Point 1, erection of a total of eight new buildings in order to shelter the shafts on top of the service experimental caverns, including all services required for the running of the ATLAS detector.
- at Point 5, construction of a total of nine new buildings to shelter the shaft on top of the experimental cavern, the single shaft on top of the service cavern, and all services required for the running of the CMS detector. The SX building was completed very early in the project to allow CMS to start pre-assembly of the detector.

Apart from the surface buildings mentioned here above, new ones have been constructed at the Points 2, 4, 6 and 8 to house the cryogenic compressors for the LHC machine. Three other steel buildings were erected at Points 2, 4 and 8 to house cryogenic equipment such as cold boxes.

Safety and the environment

CERN is proud of its record in safety and environmental protections; indeed, CERN has exercised particular vigilance in this area, mindful of the European directive emphasizing client responsibility for worksite safety and aware of the major risk potential inherent in any underground work.

For this reason the decision was taken to use French regulations (as had previously been done for the construction of the LEP), which were the most

29. Two aerial views of the major surface construction sites; (a) the ATLAS site at Point 1; and (b) the CMS site at Point 5.

30. View of the metal frame of the SMX building under construction at the CMS construction site.

advanced at the time of the call for tenders. Under this unusual arrangement, the competent Swiss authorities in this field graciously agreed to take part in this process. Accordingly, a fully-fledged cross-border health and safety organization was set up. CERN awarded the task of coordinating health and safety for the program to two companies: one for the design phase and a second for the construction phase.

The numerous worksites experienced several minor incidents, most of them relatively harmless. Indeed, no serious accident occurred, despite the magnitude and complexity of the work involved; this is a testament to the effectiveness of the prevention program.

CERN has also exercised great vigilance in the field of environmental protection, especially for the abatement of water, dust and noise pollution, in particular from the underground construction sites, where work went on continuously, with three shifts operating five or six days per week. CERN has always been very attentive to any complaints from those living in the vicinity of the work sites, on occasion going so far as to set up regular meetings with them to allow complaints to be aired and provide an opportunity for information exchange.

In particular, CERN ensured that the 600,000 m^3 of excavated rock and soil were deposited, either on the same site, or in the immediate vicinity of each site, thereby minimizing the impact of trucking traffic on the local road network.

Another remarkable illustration of this environmental concern can be seen in the arrangements made for disposing of the spoil excavated from the ATLAS area on Swiss territory. Following a year's negotiation, a temporary cross-border road was constructed so that the excavated soil and rock could be transported and deposited underneath a 400 kilovolt power line on French territory. Once all of the excavated material had been deposited, the area was re-landscaped, as was done in the case of the LEP construction sites, so that it blends into the lush surrounding landscape.

Conclusions of success

By mid 2005 all major civil engineering works for this ambitious project were complete. The inherent difficulties encountered during construction of a project of this scale within the molasse of the Lemanic basin and in the environmental constraints of the Geneva – Pays de Gex area, have been overcome without the need for any arbitration process. The success of the operation can be attributed to the experience of CERN with its previous construction sites, very careful planning, and resourceful engineering when confronted with the unexpected. The organization and technologies deployed for construction have played a major role in its success, clearing the way for the major advances it is hoped to achieve in this field of science over the next two decades.

31. Landscaping with the disposed spoil of the LHC construction sites, used to reinforce a centuries-old roman path in the vicinity of Cessy.

The construction of the LHC

4.1

THE TECHNOLOGY OF THE LHC

Superconducting Magnets

Lucio Rossi and Ezio Todesco

Superconductivity

As in other chapters of this book, our discussion must begin with a quick overview of the remarkable property of superconductivity. Some materials lose all electrical resistance below a temperature T_c, called the *critical temperature*. The lack of resistance means that these materials can carry electrical current without power dissipation. This discovery has been made possible by the first liquefaction of helium (1908) at the temperature of 4.2 K, i.e. −269°C. In 1911, H. Kamerlingh Onnes (pictured in Fig. 1) discovered in Leiden that a very pure mercury sample becomes superconductive at 4.2 K [1]; the experimental curve he measured is shown at the start of Chapter 1 (Fig. 2). Superconductivity immediately raised great hopes to produce "supercables", "supermagnets" and other powerful devices. However, the first trials were quite disappointing in terms of practical applications, and this remained the case for about 50 years. Superconductivity was long time confined to fundamental research and laboratory experiments. In 1933 Meissner and Ochsenfeld [2] discovered what today is considered as the most peculiar signature of superconductivity: an almost perfect diamagnetism, i.e. the expulsion of the magnetic flux from a superconducting specimen.

Basics physics of superconductors and critical surface
After many phenomenological theories [3-5], the mystery on the origin of superconductivity was finally unveiled about 50 years after its discovery, with the BCS theory [6], named after J. Bardeen, L. N. Cooper and J. R. Schrieffer. According to the BCS theory, superconductivity is based on coupling of electrons near the Fermi surface; under particular conditions and when brought below a transition temperature, spin-opposite electrons (particles with spin $1/2$) "condense" in pairs, forming a boson entity (i.e., a particle with integral spin) that is no longer scattered by the lattice. Therefore, the electron pairs may be accelerated and travel without resistance across meters, or even kilometers, in the superconducting material, keeping perfect phase correlation of their wave function. This is one of the most striking macroscopic manifestations of quantum mechanics.

However, the binding energy of the electron pairs is very weak – on the order of a few milli-electron volts (meV). The temperature at which the associated energy is sufficient to break the pair, the critical temperature, is very low: typically 4 to 20 K for what are known as classical superconductors (LTS for Low Temperature Superconductors), and 50-150 K for what are

1. The discoverer of the experimental evidence of superconductivity in Hg [1], H. Kamerlingh Onnes.

known as HTS (High Temperature Superconductors [7]). A second effect is that the presence of a magnetic field lowers the transition temperatures; moreover, beyond a certain magnetic field (called critical field) the material is no longer superconducting. Thirdly, the current density flowing in the superconductor also lowers the transition temperatures. Therefore, each superconductor is characterized by its critical temperature, its critical magnetic field, and its critical current density.

In the space (T,B,J) above the critical surface, the material is in the normal, resistive state, whereas below the critical surface the material is superconductive (Fig. 2, left). If we take a section of the critical surface at zero current, we have the critical field, i.e., the maximal magnetic field under which the material remains superconductive, as a function of temperature (Fig. 2, right). On the other hand, if we take a section at a fixed temperature, which is the normal operating modality for magnets, we have the critical current as a function of the field, as shown in Figure 3.

Critical current and stability

There are thousands of superconducting materials both for LTS and HTS. Many pure elements are superconducting, and so are many different alloys, metallic compounds and oxides that are nearly insulators in the normal state. Even though the first discovery of a superconductor element dates back nearly one century, new superconducting materials are being discovered every year. The most significant recent discoveries include MgB_2 [8], one of the most promising for applications, and the family of the ferro-oxypnictides, discovered in 2008 [9].

Superconductivity is not a rare phenomenon, and it still provides surprises and thrills to the scientific community. However, very few of these materials have properties good enough for practical applications: materials with critical current density in excess of 1000 A/mm^2 and critical field beyond 5-10 tesla were found only at the beginning of the 1960s, i.e., fifty years after the discovery of superconductivity. For applications, we need materials with good basic physical properties; however, materials also need to be relatively common (i.e., reasonably cheap), easy to form into the practical shapes of tape or round wire of sufficiently long length and, for most applications, easily assembled in form of cable. In addition, they have to be able to withstand me-

2. Superconductor critical surface in the space temperature, current density, magnetic field for Nb-Ti and Nb$_3$Sn (left), and critical field B$_{c2}$ vs. temperatures (at zero current density) for the most popular superconductors (with // we mean that field is applied parallel to the wide size of ribbon-shaped sample of the material).

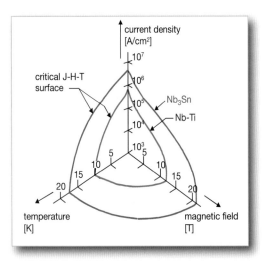

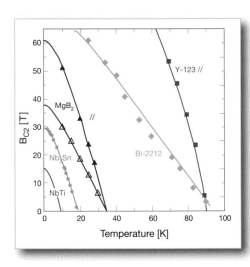

chanical stresses. The materials shown in Figure 3 are all good candidates for applications.

Both the critical temperature and the critical field are a characteristic of the material; therefore, for a given material, all experimental and industrial effort aims at improving the critical current J_c. However, improving J_c is not the whole story. The transient regimes induced by changing the current, the electromagnetic stresses and other phenomena can generate a sudden transition to resistive state. Superconductors are neither stable nor usable without embedding them in a stabilizing material that can take the current if a part of the superconductor ceases to behave as such [10]. For classical superconductors, this role is played by copper, whereas silver is used for HTS. But this is still not enough; the superconducting element must also be divided in extremely fine filaments, of 5-10 μm diameter for accelerators, and 30-70 μm for detector magnets or other applications. The small filament sizes are necessary to improve stability and to limit the undesired effect of persistent current.

The copper or silver stabilizer plays two different roles: it provides the necessary stabilization against small perturbations and "protects" the magnet when, due a too large perturbation, the material loses the superconductive state (we call this a quench). In this second case, the stabilizer must carry the enormous current density that usually flows in the superconductor, i.e., 100 to 1000 A/mm². These values are about 100 times the usual current density in copper cables, and therefore the current must be quickly damped to avoid over-voltages and the possible melting of the cable. To be effective both for stabilization and for protection at low temperatures, copper/silver must be pure in order to have good electric and thermal conductivity.

In order to have sufficient margin of stability against perturbation, a superconductor cannot operate too close to the critical surface. In general, the LHC magnets work in a range between 80-85% of the maximum current attainable, i.e. with a margin of 15-20%.

For high-temperature superconductors, stability is not critical, since they usually operate in the 10-70 K temperature range where the specific heat (i.e., the quantity of heat needed to raise the temperature of a unit mass of material by one degree) is much larger than at 2-4 K; however, the main issue for their use in practical applications is the lack of sufficient overall current density in

3. Current density averaged on the whole conductor (J_c, i.e., engineering current density) versus magnetic field for some typical superconductors at 4.2 K; also shown are the 1.9 K valves for Nb-Ti, courtesy of Prof. P. Lee, National High Field Magnet Laboratory, Tallahassee (FL), US.

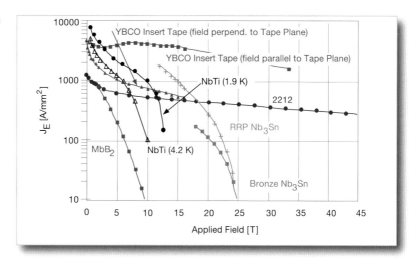

long lengths and the lack of mechanical robustness. The cost issue, especially for bismuth-based compounds, is very relevant, too.

Superconducting wires and cables for the LHC magnets
In practice, the ductile alloy niobium-titanium (Nb-Ti, 47 wt.%) and the intermetallic compound Nb_3Sn are currently the only materials used for magnets, despite the requirement that they must be operated at liquid-helium temperatures. Nb_3Sn has much better properties in term of J_c and B_c, but it is mechanically brittle and at least five times more expensive than Nb-Ti. More than 95% of the superconducting magnets built every year (about 3000, 80% of which are for magnetic resonance imaging) are wound with Nb-Ti, as are the 1700 large magnets and the 8000 corrector magnets of the LHC.

The basic element of the LHC magnets [11, 12] is the single wire, about 1 mm in diameter and composed of one third superconducting material and two thirds copper. The Nb-Ti filaments are 6-7 μm in diameter and precisely positioned with 1 μm spacing in the copper matrix. The wire is obtained by multiple co-extrusion of Nb-Ti ingots with pure copper rods and cans. The strands and the multi-strand cables are shown in Figure 4. In Figure 5 we show the distribution of the critical current measured for one type of strand and on the corresponding cables used for the LHC.

4. LHC superconductors: wire cross section (top right), zoom on the filaments (top left), cable cross-section (left, center) and cable flat side view (bottom).

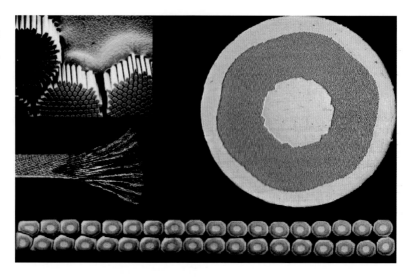

5. Critical current as measured on the entire production of the LHC-dipole inner strands (left); and for the corresponding inner cable (right) in one producer (courtesy of T. Boutboul, CERN).

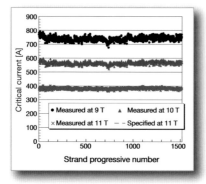

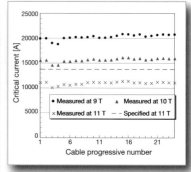

6. Two views of the 13-kA high-temperature superconducting leads used in the LHC: full lead (a) and detail (b) showing a close-up of the lower part. In (b), the transition between the high-temperature superconductor (on the left) with the low-temperature superconductor (on the right) is visible. Resting in liquid helium, the low-temperature superconductor is connected to the bus-bars conveying the current to the LHC magnets. (courtesy of A. Ballarino, CERN).

Superconducting tape for the LHC feed-through

In 1986, G. Bednorz and A. Muller discovered that the lanthanum-based cuprate compounds have a superconducting transition above 30 K [7]. They understood this to be a new class of superconductors, opening a new way towards superconductivity at "high" temperatures. Critical temperatures of 85 and 110 K (i.e., well above the 77 K boiling point of nitrogen) were quickly found for new cuprates based on yttrium and bismuth. The two scientists were awarded a Nobel Prizes that same year, one of the fastest in Nobel-Prize history. This was the fourth of the five Nobel Prize awarded for discoveries related to superconductivity, together with Onnes in 1913; Bardeen, Cooper, and Schrieffer in 1972; Esaki, Giaever and Josephson in 1973; and Abrikosov, Ginzburg and Leggett in 2003.

Among the high temperature superconductors, the material most widely used is the compound $Bi_2Sr_2Ca_2Cu_3O_{10}$ known as Bi-2223, usually with Pb doping. Its use for magnets is strongly limited by its poor critical current density; in magnetic fields, J_c is large enough only near liquid-He temperature, where classical superconductors are already available. Moreover Bi-2223 is brittle, due to its ceramic nature. The third bad feature is the cost, which is about 20 to 50 times greater than that of Nb-Ti, partially due to the requirement that silver be used as the stabilizing material.

However, at zero magnetic field, Bi-2223 can carry more than 100 A/mm^2 in a thin ribbon of 0.25×4 mm^2 at 77 K (liquid nitrogen temperature), and J_c increases at lower temperatures. These features provide a unique opportunity for using Bi-2223 as the current-lead material in the warm-to-cold transition of the current feeder (the system that brings the current from the electrical network to the magnets). At the warm end of the current lead, a large copper section (~2000 mm^2) would have been required to carry the elevated currents (13 kA) planned for the LHC; the presence of this much copper would constitute a major source of cryogenic loss in the ring due to the heat transferred to the cold superconductors by the room-temperature copper cables. This situation is much improved by intercepting the current with lighter HTSs near 60 K, because the heat input resulting from the 60-to-4-K transition can be made negligible. In this way the heat input at 4 K can be reduced by a factor of ten with respect to traditional leads such as copper. In the LHC we have to feed many circuits with total amperage of 1.5 MA; the saving is about 3 kW at 4.2 K, i.e., a non-negligible fraction of all cryogenic power. Current leads are one of the most successful applications of HTSs. The 13-kA LHC current leads are shown in Figure 6.

Design of the LHC magnets

Functions and categories

Most of the LHC does not accelerate beams, but rather guides them from the exit back to the entrance of the accelerating structure [13]. The bending is provided by vertical dipolar magnetic fields that act via the electromagnetic force. Let N be the number of dipoles and l their magnetic length: the field B determines the energy E according to:

$$E[\text{GeV}] = 0.3 \frac{N}{2\pi} \cdot l[\text{m}] \cdot B[\text{T}]$$

The total length Nl is limited by the budget of the project (and in this case the size of the existing tunnel), since it is related to the size of the machine. The

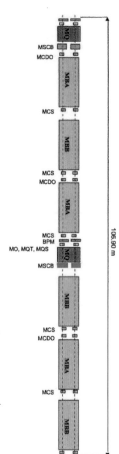

7. Sketch of the LHC main cell (MBA, MBB are dipoles; MQ are quadrupoles). An LHC arc, which takes up about 1/8 of the whole machine, is made of approximately 23 cells.

dipole length l should be chosen as long as possible; indeed, two 5-m-long magnets cost more than one 10-m-long magnet. The physical limit that determined the final value ($l = 14.3$ m, more than 15 m of physical length) was decided by the maximal length allowed by regular transport on European roads, as well as by constraints in installation in the tunnel. On the other hand, the maximum operational field was fixed at 8.3 T to have sufficient margin (for stability) with respect to the critical surface of the Nb-Ti, as we will see in more detail below. We have shown in the previous section that this second choice has its roots in the realm of quantum mechanics rather than in European Union regulations.

The LHC is divided in eight arcs separated by straight sections. In the LHC, the dipole field covers about 85% of the arcs, namely 17.6 of the 20.7 km, corresponding to $N = 1232$ dipoles, $l = 14.3$ m in length, with $B = 8.3$ T field for a 7 TeV energy. A machine consisting of only dipoles cannot function because the particles would diverge from the nominal orbit, and would be rapidly lost. A force is needed to bring them back, with a spring-like action, i.e., with a force that is zero on the reference orbit but that linearly increases with the distance from it. This force is provided by the field of the main quadrupoles (MQ).

The spacing between quadrupole magnets must be carefully considered; at larger spacing, fewer quadrupoles are needed, and thus a larger fraction of the arc becomes available for dipoles. On the other hand, greater quadrupole spacing increases the beam size, requiring a larger magnet aperture. The optimal spacing for the LHC has been set to approximately 50 m (Fig. 7), resulting in a magnet aperture of 56 mm. This has been accomplished with Nb-Ti technology providing a 223 T/m gradient over 3.15-m-long quadrupole field. The quadrupole coils take up about 6% of the arcs; and together with the dipole coils, more than 90% of the arcs are filled. The rest of the space is needed for magnet ends, corrector magnets and interconnections between adjacent magnets. Corrector magnets and higher order harmonic magnets (up to dodecapoles in the LHC) are necessary (Fig. 7) to guarantee the full beam stability, to compensate imperfections of the main magnets, and to provide flexibility to the machine operation.

Close to the experimental stations, dipoles are needed to bring the beams together at the collision points and then to separate them back to the nominal distance of 194 mm. Some of these dipoles are superconducting, built as a special contribution from the USA, while the others are resistive and have been provided by the Russian Federation. Also, in the long straight sections the optical quality needs to be preserved. This is obtained through individually powered quadrupoles, which are a mix of the standard MQ, and of quadrupoles with a different design (MQM) or a larger aperture (MQY). Finally, a large aperture triplet of quadrupoles (MQX) is placed on each side of the interaction points to be able to squeeze the beam as much as possible in order to maximize the probability of collisions between particles. These magnets have been built by the USA and Japan as a special contribution. Otherwise all LHC superconducting magnets (except some correctors) have been manufactured by European industry.

Basic design of superconducting dipoles
In superconducting magnets, most of the field is produced by a high current density flowing in a small winding. For a dipole, the coil is arranged around the beam tube according to the geometry shown in Figure 8 (top). The cross-

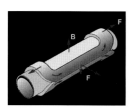

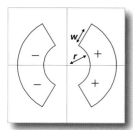

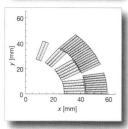

8. The 3D sketch of a cosq coil (top), and schematic cross-section of a sector coil with aperture radius r and coil width w (center), and schematic cross-section of the LHC dipole (bottom).

9. Maximum field versus coil width for five dipoles (markers), an estimate based on a simplified ironless model at 4.2 K (red line) and at 1.9 K (blue line). The same Nb-Ti cable properties have been assumed for all dipoles.

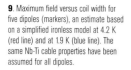

section of the winding can be approximated by a sector of thickness w and 120° angular width carrying a current density j (seen in Fig. 8 center and bottom):

$$B \sim \frac{\sqrt{3}}{\pi} \, jw \qquad B\,[\mathrm{T}] \sim 7 \times 10^{-4} \, j\,[\mathrm{A/mm^2}]\, w\,[\mathrm{mm}]$$

For the LHC main dipole, the coil thickness is about 30 mm; an average current density of 400 A/mm² provides a field of approximately 8 tesla.

The dependence of the maximum field (without the operational margin) on the coil width is shown for a series of accelerators in Figure 9. Previous coil widths ranged from 10 to 25 mm, providing fields from 4 to 6 tesla. Operation at 1.9 K instead of at 4.2 K, as in previous accelerators, results in a gain of 1-2 tesla. Doubling the coil width, i.e., from 30 to 60 mm, would only have resulted in a gain of 1 tesla, i.e., a field increase of 10%. Because the superconductor is expensive (25% of the total cost of the magnets), 10 T can be considered the ultimate limit of the Nb-Ti technology when cost is factored into the calculation.

As outlined in the previous section, the superconductor has to work at a safe distance below the critical surface; in general, one assumes a working point at 70-85% of the material's limit. For the LHC dipoles, the upper value of 85% has been retained, resulting in an operational field of 8.3 tesla [11-14]. Short models and pre-series magnets have demonstrated that the design has the potential to reach 9 T [11].

During magnet powering, strong forces push the coil outwards and towards the midplane. A mechanical structure called collars is used to keep the coil in the correct position and to avoid movement during the powering of the magnets. Collars were first introduced in the Tevatron magnets at the end of the 1970s [14]: the LHC dipoles have the special feature of having a unique collar structure for the two coils used for the oppositely rotating beams, thus providing both magnetic and mechanical coupling between the two apertures (Fig. 10). This design choice increased complexity, but in exchange provided significant cost saving (about 15%) and a more compact magnet that could fit the tight constraints of the tunnel size.

To limit the stray field, thus protecting proximity electronics and increasing the central field, the coils are surrounded by a cylinder of iron, called the yoke. Iron (actually low carbon steel) saturates just above 2 T; it can be calculated that, for the LHC dipoles 10 cm of yoke are enough to cancel any residual magnetic field outside the magnet. The iron yoke increases the maximum field by 3% only; however it reduces the operational current by about 20%, thus easing aspects related to protection and powering.

In the LHC dipoles, the iron yoke and the shrinking cylinder play a limited role in the coil mechanical support, which is mostly guaranteed by the collars.

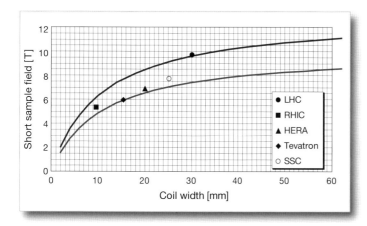

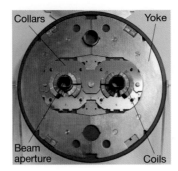

Collars   Yoke

Beam
aperture   Coils

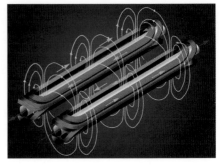

10. Cross-sectional model of the LHC dipole without cryostat (cold mass), highlighting the position of the superconducting coils with respect to the beam apertures. The drawing on the right shows the magnetic coupling between the two apertures. A detailed view of the dipole in its cryostat is shown in Fig. 15 of Chap. 1.

A 10-mm-thick stainless-steel shrinking cylinder, welded around the iron yoke, provides a vessel for the liquid helium, and gives the needed flexural rigidity. Each one of the 1232 dipoles of the machine bends the beam by $2\pi/1232$, or 0.29 degrees. Over the length of 14.3 m, this corresponds to a *sagitta* (deviation) of 9 mm. This means that if the dipoles were straight, 9 out of the 28 mm of aperture with "good field" would be lost due to the beam curvature. To avoid this waste of precious aperture, the dipoles are curved.

The main LHC quadrupoles
Quadrupole magnets are characterized by the field gradient (T/m) rather than by the field; the field gradient interacts with the off-center particles to provide the main corrective force to keep the beams tight. The main guideline was to reuse the technology, superconducting cables and components of the main dipoles to save cost. The outer-layer cable of the LHC main dipole has been used in a two-layer coil configuration, where the outer layer is wound on the top of the inner layer. This total coil width of about 30 mm provides a nominal gradient of 223 T/m over an aperture of 56 mm. This corresponds to a field on the superconductor of 6.85 T; the quadrupoles work with a larger margin than the dipoles. The collars support mechanically all electromagnetic forces, and the iron is merely a flux return yoke playing no role in the mechanical structure.

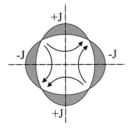

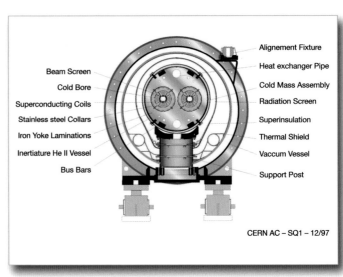

Beam Screen
Cold Bore
Superconducting Coils
Stainless steel Collars
Iron Yoke Laminations
Inertiature He II Vessel
Bus Bars

Alignement Fixture
Heat exchanger Pipe
Cold Mass Assembly
Radiation Screen
Superinsulation
Thermal Shield
Vaccum Vessel
Support Post

CERN AC – SQ1 – 12/97

11. Cross-section of a basic quadrupole field shape (above) and schematic drawing of an actual LHC quadrupoles (right); the coil is structured as a double layer.

Correctors

As mentioned above, the corrector magnets provide fine-tuning and flexibility in beam manipulation; at the LHC, we have sextupoles, octupoles, decupoles and dodecupoles. All the correctors are superconducting single-bore modules, working with a large operational margin (40-60%). The coil is made up of flat multi-wire ribbons of Nb-Ti strands with diameters ranging from 0.3 to 1 mm. The wires of the same cable are then connected in series at the end, allowing a low operational current, a feature that is important to simplify powering (there are 8000 corrector magnets in the LHC). A special iron-yoke structure, called "scissor laminations", provides the mechanical support for retaining the forces acting on the coil, while at the same time providing a field enhancement. The final assembly of the MCS (sextupolar corrector in the dipoles) is shown in Figure 12.

Development of the main components

Porous insulation and collars

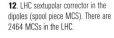

12. LHC sextupolar corrector in the dipoles (spool piece MCS). There are 2464 MCSs in the LHC.

When a *quench* occurs, i.e., when the superconductor returns to its normal conducting phase, the resulting current flowing through the copper creates a turn-to-turn voltage on the order of 20 V. In addition, the voltage at the terminals and the voltage to ground may rise to 500-1000 V. The insulation has to withstand this voltage, but at the same time must be porous to liquid helium to allow effective cooling of the coils. The adopted solution is to insulate the cable by wrapping three layers of polyimide tape. Each layer has a thickness of about 0.05 mm; the first two layers have a 50% overlap to avoid any free path between the adjacent cables. The third layer is wrapped with a 5-mm gap (Fig. 13) to leave micro-channels between cables, thus allowing the superfluid helium to penetrate the coil and to have a more efficient heat removal.

The mechanical structure used to restrain the coil and to withstand the large electromagnetic forces arising during magnet powering is based on austenitic steel collars common to both apertures (Fig. 14). Collars are not machined in a single piece but they are made of packs of 3-mm-thick laminations. Since stainless steel is paramagnetic, it affects the field homogeneity. For this reason, the magnetic properties of the collar have to satisfy tight toler-

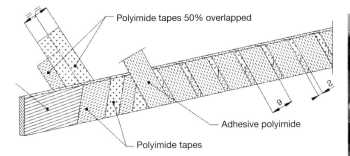

Polyimide tapes 50% overlapped

Adhesive polyimide

Polyimide tapes

13. Insulation scheme of the LHC cable (left) and wrapping of the third layer (right).

ances, and a low-permeability stainless steel has been chosen. Mechanical tolerances on the collar shape are ± 0.02 to ± 0.03 mm; this is particularly challenging in the LHC case, where the shape of the twin collars is extremely complex (Fig. 14).

Protection components: quench heaters and diodes

The energy stored in the magnetic field of all the LHC dipoles under operational conditions is one order of magnitude larger than the energy of the beams. For the twin aperture dipole operating at 8.3 T, the stored energy is 7 MJ; the energy of the string of 154 dipoles – powered in series in each octant – is about 1 GJ. This is the kinetic energy of a fully loaded 200-ton Boeing 767 approaching the tarmac at 350 km/h, or the energy required to lift the Eiffel tower by 10 meters. In case of a transition from super to normal conducting state – the quench – this energy has to be safely removed from the magnets [10,12-14].

The first issue of magnet protection is to avoid excessive energy deposition in the tiny zone (of the order of a few millimeters) where the transition to the normal state occurs. As soon as the transition is detected, the energy dissipation is spread over the largest possible area. This is done by bringing the whole coil in a normal conducting state by heating it with "quench heaters", placed between the coils and the collars, capable of raising the coil temperature by 10 K in 25 ms.

The second stage of magnet protection then kicks in to prevent the energy of the other 153 dipoles of the octant from being dissipated in the quenching dipole. Each dipole, and any large magnet, features a high-current bypass diode operating at 1.9 K. When the quench occurs, the resistance of the coil induces a voltage that opens the diode, and the current through the quenching dipole decays in less than one second. When the quench is detected, the power supply is switched off, and the diode conducts an initial current of 12 kA, with a decay time of around 100 s.

Cryostats and interface

Just as the yoke shields the devices located in the tunnel from the magnetic field inside the magnet, the cryostat shields the coils operating at 1.9 K from room temperature heating. When the coils, the collars and the yoke are at 1.9 K, the external part of the cryostat is at room temperature and only about 20 cm separate materials at these extreme temperatures, corresponding to a gradient of approximately 10 °C/cm.

The helium vessel, formed by the shrinking cylinder and by the end caps, is the primary element of the cryostat. To reduce heat transfer by conduction and convection it is housed in a second cylinder kept under a vacuum of about 10^{-9} atm, called the vacuum vessel. Its size is exactly that of a standard oil pipe, to profit from large-scale industry production. It is made of carbon steel, to both minimize costs and to provide a further magnetic screen to avoid stray fields in the tunnel. An intermediate aluminum cylinder, kept at approximately 60 K, intercepts the radiation heat from the room-temperature surroundings; since radiation power scales with T^4, the presence of this thermal shield reduces the radiation power onto the helium vessel by a factor of 600. What remains is the power radiated by the thermal shield and the conduction through the posts that support the 30-ton cold masses. The three support posts necessary to support the static and the dynamic load (20% greater than the gravitational force) are of a rather complex design. They are made of composite material to provide the necessary stiffness, while minimizing the conduction heat input: the heat loss of each support post is only 0.05 W at 1.9 K and 0.5 W at 10 K, while 6 W are intercepted at 50 K. This engineering challenge is similar to that of making special gloves that allow you to hold for long time a very heavy block of ice without freezing your fingers! A view of the dipole support post is given much later in this Chapter (Fig. 51).

In addition to providing the access for the instrumentation for each magnet, the cryostat is the interface of the magnet cold mass with the floor, and therefore it is of primary importance for the geometry. A longitudinal sliding of the two lateral support posts is necessary to accommodate the 5-cm contraction of the cold mass with respect to the room-temperature vacuum vessel when passing from 300 K to liquid-helium temperature. The sliding of the central support post is important for the final shape of the cold mass, which has to follow precisely the 9-mm sagitta to within 1 mm and even 0.3 mm at the extremities. The accuracy in obtaining the geometry has been one of the practical challenges in the main dipole production. In order to precisely measure the curvature shape inside the 15-m-long dipole and to position the cold mass inside the cryostat at ± 0.2 mm, a survey system based on laser tracking had to be developed. It is worth mentioning that the cryostat for the short straight sections housing the quadrupoles (SSS cryostats) is far more complex than the dipole cryostat because the SSS has to accommodate 65 different magnet types!

15. The laser tracker developed at CERN to measure the magnet shape, with its accompanying instrumentation.

Construction

The LHC dipole cold mass construction

The LHC dipole cold masses, i.e. the part that, during operation, has the same temperature as the superconducting cable, were assembled by three different manufacturers, with main components provided by CERN. The assembly operation is about one third of the total cost of the magnet and involves extensive dedicated tooling. The first operation is the highly precise winding of the insulated cables; the coil is then closed using a mold that when under pressure reaches the nominal size; then it is heated to activate the glue located between the turns of the insulation. With this procedure, the coil position can be controlled with a reproducibility of about 0.03 mm. This level of precision in the coil position (along 15 m!) is needed to obtain the required homogeneity of magnetic field. After curing, the collars are assembled around the four poles that constitute the two dipoles. The whole assembly is then inserted into a 15-m-long press that locks the collar keys (collaring) to give the necessary pre-compression in the coil midplane (Fig. 16, left).

The collaring concludes one of the most delicate phases, i.e. the construction of the active part of the magnet: the coils have now reached their nominal position in room temperature conditions and are no longer accessible. A low current (~10 A) is fed through the cables, and magnetic measurements are done to verify the electrical integrity and the correct cable position and to obtain a first assessment of the field homogeneity.

The iron yoke and the busbars are then assembled around the collared coil, and the two half-shells are welded around the assembly with another dedicated, 15-m-long press designed to introduce a circumferential stress of about 150 MPa (Fig. 16, right). This operation provides the final longitudinal shape to the dipole, i.e. an arc with a sagitta of 9 mm. The assembly of spool-piece correctors on the magnet ends, and the welding of the end covers completes the construction of the cold mass. These are then shipped to CERN by road.

QA, non-conformities, and analysis of the industry production

One of the main challenges of the LHC dipoles is to have the whole set of 1232 dipoles working at 85% of the theoretical limit of the superconductor performance: this requires careful overseeing of the production, with intermediate tests and procedures for rejecting faulty parts. Considering the total cost of a single dipole (about one million Swiss francs), the quality control cannot

16. LHC coils ready for the collaring (left) and cold mass ready for welding (right).

17. A case of bad coil curing resulting in an inner radial movement of two turns of the inner layer, upper pole, of about 1 mm (left); and a misplaced folded shim (green strip in the right part), found through anomalies in room temperature magnetic measurements.

rely on the rejection of the final product, and all tools are designed to intercept a fault in the production chain as soon as possible.

Numerous series of tests are conducted on components – starting with the most costly one, the superconducting cable – and partial assemblies, mainly dimensional measurements, electrical tests and checks of the magnetic properties. At the manufacturer, magnetic measurements at room temperature have been used as a sort of "X-ray machine" to detect anomalies resulting from the assembly procedure. In all, 19 faulty assemblies (1.6% of total) were intercepted, disassembled and repaired (Fig. 17).

The final acceptance tests were conducted at CERN after placement of the magnet in the cryostat. The dipoles and quadrupoles were tested in operational conditions at 1.9 K. In all, 31 magnets (2.4% of the production) were returned to the manufacturer after cold tests: 14 for insufficient quench performance and 10 for electrical shorts or insulation faults. With the exception of one case, all defects were reparable.

Performance

Cold powering and quench performance

All the LHC main magnets were "cold tested" under operational conditions at CERN in a dedicated test station (Fig. 18). Cold tests are designed to verify the electrical and mechanical integrity at 1.9 K, as well as to assess the performance in terms of maximum attainable magnetic field (quench performance). The level of the first quench in a superconducting magnet is highly variable, even within magnets built with the same design and by the same manufacturer. Typically one observes first quench at 70-90% from the critical surface. For the large production of the LHC case, a first quench value of 70-90%

18. LHC dipoles undergoing reception tests at cryogenic test benches at CERN.

(i.e., for magnetic fields 7 to 9 tesla) was obtained for 95% of the dipoles, and the remaining 5% had a first quench above 60% (see distribution in Fig. 19).

If the magnet is not limited by a degradation of the superconductor, by instabilities, or by spot defects, the magnet trains, i.e., it has successive quenches at successively higher fields to approach the critical surface (Fig. 20). As shown in Figure 19, the second quench level is in average 0.5 T larger than the first one. Training quenches are due to the release of extremely small amounts of energy (10-100 nJ) during coil movements, pushed by the electromagnetic forces, and subject to friction. Even though the physical principles governing the training are well understood in principle, and notwithstanding the experience acquired in the production of a few thousand superconducting magnets for large accelerators, today we cannot claim to have attained a level of control of the manufacturing process that would allow us to push the magnets close to their limits without training.

But this is not the end of the story. If the magnet is warmed and then cooled down again (called a thermal cycle), the first quench normally happens at a higher level than the first virgin quench (Figs. 19 and 20). The initial strategy followed for the LHC dipoles has been to train all magnets up to 9 T to establish the ultimate potential of the machine and, especially, to ensure a safe operation at 8.3 T. During production, this ambitious goal was lowered to 8.4-8.6 T to reduce the time between the production and the installation. About 11% of the production was tested after a thermal cycle; mostly magnets that performed poorly during the first round were selected for this additional test. On this biased sample, we have estimated that the average level of the first quench after thermal cycle is similar to the second virgin quench, i.e. about 8.5 T. In 2009, the machine will be operated at 6.5 T, corresponding to 5 + 5 TeV of collision energy. Later, it will be trained to higher field energy collisions, i.e., between 6.5 TeV and the nominal 7 TeV per beam.

Field quality

The quality of the magnetic field in a superconducting magnet is mainly determined by the position of the superconducting cables. This translates into tolerances on the order of 0.03 mm for the position of the Rutherford cable in order to obtain well-behaved bulk property. Fortunately, cable position (the main mechanism governing field quality) can be monitored with room temperature measurements, where a small current of about 10 A is circulated through the stabilizing part (copper) of the superconducting cable.

The most important parameter is the magnet-to-magnet uniformity of the dipole strength, i.e. the integrated field in T × m provided for a given current. The beam dynamics of the machine require that the LHC dipoles remain

19. First and second quench of all the LHC dipoles during test at 1.9 K, and first quench after thermal cycle for the 11% sample tested.

20. Typical training of an LHC dipole, and detraining after thermal cycle.

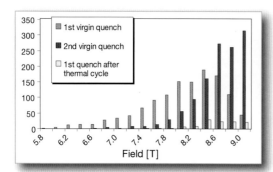

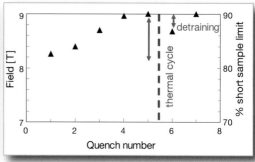

within a 0.08% r.m.s. distribution over the whole production. Just to give an idea of what this targets implies for magnet builders, this corresponds to a variation in the coil inner radius of 0.02 mm or less. Room temperature measurements over all the production (Fig. 21) have shown that this target was attained, with a remarkable homogeneity between the three magnet assemblers.

Powering and protection

Cold power

There are more than 1000 superconducting circuits that need to be powered, with an amazing number of variants; in this short discussion, we focus on the three main 13 kA circuits (one with 154 dipoles, and two with 22 or 23 main quadrupoles) present in each octant of the machine. The current for these circuits must remain exceptionally stable over long periods of time: within 20-50 ppm (parts per million, i.e., 10^{-6}) over one year and 5 ppm over one day. The stability over a half hour of constant machine settings must be within 3 ppm, and the resolution (smallest increment in current) is limited to 1 ppm. These stringent requirements are imposed by the machine operation and beam dynamics and are achieved by a new generation of suitably developed power converters.

A special cryostat, called DFB (Distribution Feed Boxes) is needed to connect the room temperature power converter to the magnets at 1.9 K. Power converters distribute current in 106-mm-diameter copper/aluminum resistive cables that are rubber insulated and water cooled. Magnets receive current from copper stabilized Nb-Ti superconducting busbars that carry the current all along the 3.5-km-long continuous cryostat of each sector. In the DFB, the entering current leads are copper cables that take the current from room temperature up to the 60 K thermal anchoring. Then the current leads made with HTS material described earlier in this chapter bring the current from 60 K down to 4 K, before connecting to the Nb-Ti cables. The leads need to be carefully monitored because their correct operation depends on many parameters and may suffer a dangerous thermal runaway if the cooling is stopped or even reduced.

Quench protection

As described earlier in this chapter, a careful strategy has to be followed to avoid that a magnet is permanently damaged during a quench. The quench detection can be schematized as a hard-wired circuit that compares voltage of different poles of the same magnet. If a certain voltage threshold is exceeded for a minimum time, a quench is assumed and a capacitor bank is automatically discharged into resistors inside the magnets (these are the quench heaters

21. Evolution of integrated transfer function measured at room temperature during the LHC dipole production, shown for the three assemblers.

22. Distribution FeedBox (DFB) in the LHC tunnel which house the high-temperature conducting leads shown in Fig. 6.

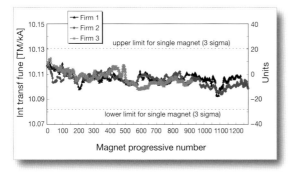

23. Image of dipole (left) being lowered into the LHC tunnel; and (right) a photo of the installed dipole with an open connection.

described earlier). This initiates a global quench that raises the magnet terminal voltage above the diode conduction threshold. For the LHC main circuits, the voltage threshold to trigger the quench heater discharge is 100 mV for at least 10 ms. Over the next 20-30 ms, heat enters in the coils and the coils start to discharge into the bus bars. With these values, we achieve a balance between different needs: on one hand we cannot risk delaying the action of firing the heaters when confronted with a real quench (another 50 ms may lead to a peak temperature in the coil of more than 400 K, with the major risk of damaging the insulation); on the other hand, frequent unnecessary quenches, resulting in magnet heating to more than 300 K within a few seconds, may become the major mechanism for the aging of the magnets.

Installation and interconnection

All 1232 LHC dipoles and most of the other 500 large quadrupoles were lowered into the LHC tunnel through the same elliptically-shaped pit; to save money on civil engineering, the pit dimensions were only slightly larger than the dipoles with no possibility to rotate the magnet to correct mistakes. The dipoles were then transported at 2 km/h on automatically guided vehicles in the narrow tunnel (the clearance point was of the order of a centimeter), and then each magnet in its cryostat is posed onto two supports anchored to the tunnel and adjustable along the three axis in order to allow for a precise alignment by reference to fiducials on the cryostat (the magnet itself no longer accessible once placed in the tunnel).

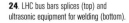

24. LHC bus bars splices (top) and ultrasonic equipment for welding (bottom).

Then, the magnets and cryostats are interconnected. The 13-kA superconducting cables of the main circuits are connected though resistive joints of typically 1 nΩ, by means of soft solder melted through an inductive heating technology, specially developed for LHC (Fig. 24). Single superconducting wires for corrector magnets are welded through ultrasonic heating, specially developed for the LHC, too. In total, about 10,000 large cables and 60,000 wires were connected in the LHC tunnel over a two-year period.

Another formidable challenge is to form the two 27-km-long beam pipes, requiring about 4000 ultra-high-vacuum welds that will function at cryogenic temperatures; in order to connect all cryogenic and vacuum pipes, about 40,000 welds were necessary in the magnet interconnections or in the cryogenic interfaces (Fig. 24).

84

In total, the interconnections form the 3.4-km-long continuous cryostat that covers almost the entire sector. Each one of the eight continuous cryostats contains about 200 large magnets, and 1000 correctors, grouped in about 15 circuits. At one point, a team of 200 people were working in the tunnel and more than 100 people were dedicated to quality control, especially for the vacuum welding and the electrical checks. In Figure 24, the ultrasonic welding equipment is shown; in Figure 25 is a photo of the ELQA (ELectrical Quality Assurance) team, ready to enter in the tunnel for routine checks.

Current and future challenges

The design, construction, installation and commissioning of the LHC superconducting magnets has been a fascinating adventure that has lasted two decades. The first challenge is the safe operation of the magnets, with respect both to the their enormous energy (some 11 GJ in total) as well as to the beam energy (350 MJ) that can easily damage many magnets or impede their reliable operation. The beam control system and the machine protection system are very sophisticated and should allow us, in a few years, to reach the nominal performance of the magnets and of the entire LHC machine.

The second challenge is the development of larger aperture quadrupoles for the interaction region of the LHC, to increase the beam collimation and therefore the collision rate. This new adventure has already started, with the goal of attaining, in a few years, the fitting of superconducting magnets capable of peak fields in the range of 15 tesla, a step requiring the development of a new class of magnets, based on superconductors such as Nb_3Sn, MgB_2 or HTS (as shown in Fig. 2).

25. Electrical Quality Assurance team in the tunnel for routine checks.

References
[1] H. Kamerlingh Onnes, *Commun. Phys. Lab. Univ. Leiden* **12** (1911) 120.
[2] W. Meissner and R. Ochsenfeld, *Naturwiss.* **21** (1933) 787.
[3] C. J. Gorter and H. B. G. Casimir, *Physik. Z.* **35** (1934) 963 and *Z. Physik* **15** (1934) 539.
[4] F. London and H. London, *Proc. Roy. Soc.* (London) **A155** (1935) 71.

[5] V. L. Ginzburg and L. D. Landau, F.E.T.P. **20** (1950) 1064.
[6] J. Bardeen, L. N. Cooper and J. R. Schrieffer, *Phys. Rev.* **108** (1957) 1175.
[7] J. G. Bednorz and K. A. Müller, *Z. Physik* **B 64** (1986) 189.
[8] J. Nagamatsu, N. Nakagawa, T. Muranaka, Y. Zenitani and J. Akimitsu, *Nature* **410** (2001) 63.
[9] Y. Kamihara, T. Watanabe,

M. Hirano, and H. Hosono, J. Am. Chem. Soc. **130** (2008) 3296.
[10] M. N. Wilson, *Superconducting Magnets* (Claredon Press, Oxford, 1983), and K. H. Mess, P. Schmuser and R. Wolff, *Superconducting Accelerator Magnets* (World Scientific, Singapore, 1996).
[11] L. Rossi, *IEEE Trans. Appl. Supercond.* **17** (2007) 1005.
[12] L. Evans, *LHC Project*

Report **635** (2003), and O. Bruning, et al., *LHC Design Report*, CERN-2004-003.
[13] R. Perin, *IEEE Trans. Appl. Supercond.* **3** (1991) 666.
[14] A. Tollestrup and E. Todesco, *Rev. Accel. Scie. Techn.* **1** (2008) 185.
[15] K. Koepke, G. Kalbfleisch, W. Hanson, A. Tollestrup, J. O'Meara and J. Saarivirta, *IEEE Trans. Magnetics* **15** (1979) 658.

4.2

THE TECHNOLOGY OF THE LHC

The LHC and its Vacuum Technology

Pierre Strubin and Cristoforo Benvenuti

State-of-the-art vacuum technology is one of the pillars of the LHC. In this chapter, we present a summary of the basic needs of the machine in terms of vacuum, before discussing in more detail three novel elements of the beam-vacuum system that allowed us to meet the performance criteria required during routine operation of LHC, taking into account undesirable dynamic phenomena. Specifically, the techniques presented involve mitigating heat sources of the circulating beam by use of a beam screen; improving a special family of chemical pumps (referred to as non-evaporable getters); and the development of the connectors that are located between adjacent magnet cryostats, called the cold interconnects, the design of which involved trade-offs permitting them to function over rather extreme ranges of temperature and pressure. This aspect of the development of LHC thus ranges from the routine installation of adequate components to the frontiers of vacuum science, with a number of brilliant successes, as well as a few setbacks that needed to be overcome along the way, thanks to the fast thinking and ingenuity of our team.

Insulation Vacuum

The LHC has the particularity of having not one, but three vacuum systems: insulation vacuum for the cryomagnets, insulation vacuum for the helium distribution line (QRL), and beam vacuum.

We consider first the insulation vacuum, one of the technologies required for the extremely high thermal insulation between the magnets at 1.9 K and the tunnel, which is at room temperature. The magnets are installed in a vessel (a cryostat), which is pumped-out (Fig. 26). Heat cannot propagate through vacuum, and so the only remaining heat sources are due to conduction via the unavoidable magnet supports and by radiation. An everyday item using the same principle is a thermos bottle in which tea or coffee can be kept warm for a long time: it is made out of a double wall with vacuum between the two walls.

Driven by the requirements of the cryogenic system, the room temperature pressure of the insulation vacuum before cool-down does not have to be better than 10 Pa (10^{-1} mbar or 10^{-4} atmospheres), a value which is relatively easy to obtain by standard mechanical pumps, even for large volumes. At cryogenic temperatures, in the absence of any significant leak, the pressure will stabilize at five orders of magnitude below this value (10^{-9} atm) because the residual gas molecules are condensed onto the cold surfaces and stick to them

The Technology of the LHC

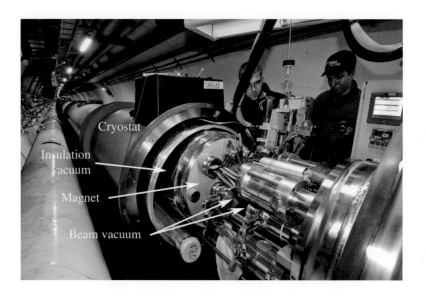

26. The open interconnection showing a magnet in its cryostat

until the temperature is raised again. If a small leak develops on the envelop of the magnet cryostat, the gas molecules coming from the tunnel air will also be condensed onto the cold surfaces. The performance of the cryogenic system will not be changed, but there will be more molecules of gas released from the surfaces when warming the system up, and the magnet enclosure could be submitted to an overpressure. It was therefore necessary to install safety devices on the insulation vacuum systems.

A more difficult situation would be a leak between a helium circuit and the insulation vacuum, because helium does not condense on the surfaces, even at very low temperatures. To cope with the eventuality of a helium leak, permanently installed turbo-molecular pumps will be automatically started to limit the pressure rise to an acceptable value until a proper repair can be made. In summary, the insulation vacuum does not require any new technology; it relies on a combination of standard pumps to start-up and the surfaces at the temperature of the liquid helium, leading to a very efficient cryogenic pumping system. Not to be forgotten, the performance can only be guaranteed if the whole system is free from any significant leak.

Beam vacuum

The beam vacuum is required to minimize the probability that a proton in the LHC hits a gas molecule and gets lost while it is going round the 27-km circumference some 10,000 times each second, hence maximizing the useful beam lifetime. The interaction between the protons in the accelerator and the rest gas is commonly referred to as *beam-gas scattering* and has a strong impact on the lifetime of the beam. The beam-gas scattering is driven by two processes, both of which are proportional to the residual pressure in the beam pipe: (1) single proton-nuclear collisions leading to a high-energy collision in which a proton is lost and (2) multiple small-angle proton-nuclear Coulomb collisions in which the beam protons are deflected, leading to an increase in *emittance* (i.e., a loss of beam collimation). The first process is not only undesirable because the number of particles for physics is reduced, but also because the lost particles will activate the materials they are impinging. Not only must the residual pressure be minimized, but the gas composition must also

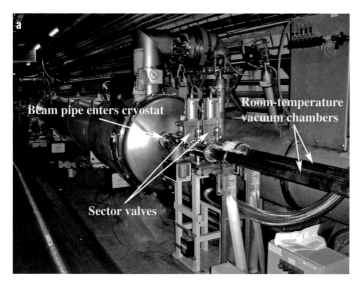

Beam pipe enters cryostat

Room-temperature vacuum chambers

Sector valves

27. (a) Sector valves at a cold to warm transition; (b) a close-up view of the sector valves.

be controlled, as the heavier molecules like CO_2 have a larger probability to interact than the lighter ones like H_2.

There are also a number of dynamic phenomena to be taken into account during the design of the beam-vacuum system. Synchrotron radiation will hit the vacuum chambers mainly in the arc, and electron clouds (*multipacting*) could affect almost the entire ring. Specific equipment has been added in the cold part, like the beam screen (described later in this chapter) to minimize the impact of these dynamic phenomena with respect to the pressure and the heat transferred to the cold magnets.

The required pressure is therefore much lower for the beam vacuum than for the insulation vacuum and falls in the range of Ultra High Vacuum (UHV), with values as low as 10^{-14} atm, required to ensure 100 hours of beam lifetime per run. In the cold parts, the pumping is insured by the cold surfaces for all gases but helium, as for the insulation vacuum. However, to minimize the desorbing of condensed gas molecules via dynamic effects, the residual pressure before cool-down should be much lower than what is required for the insulation vacuum. Sets of mobile turbo-molecular pumps, backed-up by rotary vane pumps, reduce the initial pressure to 10^{-9} atm before cooling down.

The requirements for the room-temperature part of the accelerator beam pipe are driven by the background to the experiments as well as by the beam lifetime; these requirements call for a value in the range from 10^{-8} to 10^{-9} Pa (10^{-13} to 10^{-14} atm). This high vacuum requires the system to be heated to 300 °C to eliminate the heavier gas species. Thin film getter coatings, also called Non Evaporable Getter coatings (NEG, described later in this chapter) provide most of the pumping capacity, with additional sputter ion pumps for the noble gases which are not pumped by the NEG.

The beam vacuum system is divided into sectors of manageable lengths using more than 300 all-metal valves. The most common location for a valve is at the transition between a cold and a room-temperature part (Fig. 27), so as to allow warming of the cold part without saturating the NEG coating and losing the vacuum in the adjacent room temperature parts; or to perform the bake-out of the room temperature sector (i.e., the warming of the room-

28. (a) Light shining in the non-reflective direction of the beam screen, highlighting the saw-tooth surface treatment that minimizes reflected photons; (b) light shining in the reflective direction of the same beam screen.

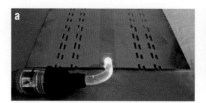 

temperature sector to remove gas and thus improve the vacuum) without loading the cold part with gas.

The performance of the beam-vacuum system will ultimately be assessed by the quality of the beam, but sets of gauges are also installed, all in the room temperature parts. A few vacuum sectors will be equipped with more instrumentation, integrating also residual-gas analyzers to assess the quality of the vacuum and injection facilities to measure the residual pumping speed of the NEG coatings.

The required leak tightness is also very challenging, with detection levels of better than 10^{-9} Pa m^3 s^{-1}. To illustrate what this level means, if a car tire would have the same level of leak rate as the beam vacuum system of LHC, it would loose half of its inflating pressure only after several million years.

Now that we have concluded the general introduction, the remaining part of this chapter will highlight three specific aspects of the LHC vacuum system: the beam screen, the NEG coatings and the interconnections in the cold part.

Heat meets vacuum: the beam screen

Any heat load into the magnets must be minimized, as it affects the performance at high field. As the beam pipe is embedded into the magnet, the vacuum team is also highly concerned about the quality of the thermal conditions of the LHC. Indeed, basic physics teaches us that there are several heat sources associated with a circulating proton beam; each of these were carefully evaluated during the design phase of LHC. The four main ones are

- synchrotron light radiated by the high energy circulating proton beams (0.2 W m^{-1} per beam, with a critical energy of about 44 eV);
- energy loss by nuclear scattering (30 mW m^{-1} per beam);
- energy dissipation by image currents (0.2 W m^{-1} per beam);
- energy dissipated during the development of electron clouds, which will form when the surfaces seen by the beams have too high a secondary electron yield.

The beam tube in the magnet, also referred to as the cold bore, is immersed in the 1.9-K helium bath of the magnet. As everything was made to minimize the heat input to the 1.9-K cryogenic circuit (Chap 4.1), a solution had to be found to intercept and evacuate the power sources mentioned above at a higher temperature, i.e., on a surface other than the beam tube. This led to a major conceptual improvement, namely the idea of an internal sleeve that shields the cold bore from the beam-generated heat loads; this is the device we refer to as the *beam screen*. Its cross-section has a racetrack shape, which optimizes the available aperture while leaving space for the cooling tubes (Fig. 29). The nominal horizontal and vertical apertures are 44.04 mm and 34.28 mm, respectively, but other sizes had to be built to adapt to different magnet apertures. The beam screen is maintained at a temperature between 5 and 20 K by circulating helium through two cooling tubes welded on the flat sides of the beam screen as seen in Figure 30, thus evacuating the heat load at a temperature higher than 1.9 K.

Beam screen design criteria

Three additional design criteria were selected to optimize performance of the beam screen. Firstly, as the cold bore at 1.9 K is an almost ideal pump for all gas species but helium, rectangular holes (referred to as pumping slots) with a total surface area of 4% are perforated in the flat parts of the beam screen to allow condensation of the gas on cold bore surface, now protected against the direct impact of energetic particles (ions, electrons and photons). The quasi-randomized pattern of the pumping slots has been chosen to minimize longitudinal and transverse impedance, and the size was selected to keep the RF losses through the holes below 1 mW m^{-1}. Secondly, its surface was produced with a saw-tooth structure to intercept the photons of the synchrotron radiation in the horizontal plane and minimize the amount of reflected photons. Figure 28 illustrates how this works: on Figure 28(a) the light shines towards the steep edge of the saw teeth (perpendicular incidence) whereas on Figure 28(b) it shines towards the sloping edge (grazing incidence). Finally, a thin copper layer (75 μm) on the inner surface of the beam screen provides a low resistance path for the image current of the beam, i.e., the current that the beam generates on the accelerator walls by electrical induction.

Manufacture of the beam screens

The history of the beam-screen design underscores the need to accommodate unforeseen phenomena, despite the most careful study and planning, and to resolve these issues through intelligent fixes.

The manufacturing process starts by co-laminating a specially developed low permeability 1-mm-thick austenitic stainless steel strip with a 75-μm high-purity copper sheet, and then rolling the saw-tooth structure which will intercept photons at normal incidence. The co-laminated strip subsequently undergoes a partial annealing treatment in a continuous furnace to restore the mechanical properties of the stainless steel and to increase the residual resistivity ratio (RRR value) of the copper layer, which defines the heat dissipated by the image current of the beam. The pumping slots are punched into this composite strip, which is then rolled into its final shape and closed by a longitudinal laser weld. The grade of the stainless steel has been specifically developed to maintain a homogeneous fully austenitic structure in the welds as well in order to minimize the perturbation of the magnetic field seen by the beam.

The beam screen is cooled by two seamless stainless steel tubes, with an inner diameter of 3.7 mm and a wall thickness of 0.53 mm, allowing for the extraction of up to 2×1.13 W per meter of beam screen in nominal cryogenic

29. The beam screen extremity with sliding ring and cooling tubes.

30. The cross section showing the position of the pumping slot shield.

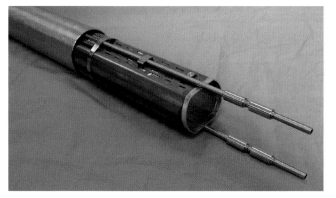

conditions. The helium temperature is regulated to 20 K at the output of the cooling circuit at every half-cell, resulting in a temperature of the cooling tubes between 5 and 20 K for nominal cryogenic conditions. The cooling tubes are laser welded onto the beam screen tube, entailing a total of 80,000,000 laser welds that all have to be leak tight to prevent the cooling-circuit helium from entering into the beam vacuum. The cooling tubes are fitted with adaptor pieces at each end, which allow their routing out of the cold bore without any fully penetrating weld between the helium circuit and the beam vacuum.

A particular concern for any stainless steel component of a vacuum system is *stress corrosion cracking*, which can be initiated by the presence of traces of halides, such as chlorides or fluorides, and accelerated by thermal cycles (e.g. welding or cool-down) and in presence of ionizing radiation (such as beam losses). Indeed, a leak on a beam screen cooling tube attributed to corrosion was discovered in the String 2 test cell built to validate the magnets between 1998 and 2004. Traces of chlorides were found inside the faulty cooling tube, probably residues of the chloride-containing lubricants used during the manufacture of the tube that had not been properly removed during the final cleaning process. After having integrated an additional nitric acid rinse before the final cleaning process at the manufacturer, all beam screens were finally cold-tested at 80 K before being installed in the cold bores of the magnets.

So-called sliding rings with a bronze layer towards the cold bore are welded every 750 mm onto the beam screen to ease its insertion into the cold bore, to ensure centering in the cold bore and to provide for a good thermal insulation. Figure 29 shows the extremity of a standard beam screen, with the last sliding ring and the cooling tubes prepared for the final assembly.

Late modifications

When it became clear that electron clouds (i.e., electrons that surround the beam because of Coulomb attraction) would appear in the LHC, and that they would be capable of depositing up to 500 mW m⁻¹ per beam into the cold bore at 1.9 K through the pumping slots, a way to shield the cold bore from the electrons had to be found. At that time, the beam screens were already manufactured and delivered to CERN. The situation called for retrofitting the existing equipment, while preserving the pumping of the gas from the beam path to the cold-bore. In addition, any solution would have to allow for the thermalisation of the shield to the beam screen cooling tubes and minimize thermal contact with the cold bore. A clever solution, consisting of preformed copper-beryllium strips, which are "clipped" onto the cooling tubes of the beam screen was proposed and implemented (Fig. 30). This solution fulfils all the requirements, in particular the one of pumping with help of additional pumping holes in the "wings" of the shields. Despite the late date of this modification request, there was no delay in the supply of fully equipped beam screens.

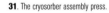

31. The cryosorber assembly press.

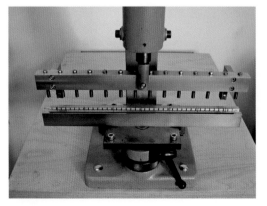

The beam screens had one last technological hurdle to overcome: some magnets in the region of the interaction points, the so-called long straight sections, are not cooled to 1.9 K, but to 4.5 K. At this temperature, the cryo-pumping capacity for hydrogen is significantly reduced, unless one uses a special pumping material, called a cryosorber.

For our application we used one made of carbon fiber in strips 370-mm long and 8-mm wide. As much as the requirement for shielding the cold bore from electrons was an additional difficulty, the proposed solution provided an elegant way to support the cryosorber strips: they are riveted between perforated support strips and the "wings" of the pumping slot shields using a mechanical press as shown in Figure 31.

To conclude this discussion, we have seen how the final, complex structure of one of the key elements depended not only on the careful planning and execution of the vacuum design team, but also evolved towards its final form for which clever technological improvisation played a non-negligible role. Table 1 summarizes the most important technical solutions adopted.

Table 1. Summary of design criteria for the beam screen.

Design options	Reason
Racetrack shape	Optimize aperture for beam while accommodating the cooling tubes
Saw tooth surface	Prevent forward scattering of photons
Choice of steel-copper co-lamination	Provide mechanical properties (steel) and low electrical resistance (copper)
Pumping slots	Allow to pump the gas from the beam path to the 1.9 K surface
Pumping slot shields	Prevent electrons accelerated by the beam to deposit energy in the magnets
Copper beryllium used for pumping slot shields	Make them act as a spring to "clip" them on the cooling tubes of the beam screen
Gold finishing of the extremity of the beam screen	Minimize electrical resistance between beams screen and interconnection module
Cryosorber	Increase pumping capacity for hydrogen in magnet operated at 4.5 K instead of 1.9 K

Thin film getter coatings

Getters are "pumps" consisting of materials (usually metallic) capable of fixing gas molecules on their surface by irreversible chemical reaction. Gas molecules are therefore converted into stable chemical compounds (surface pumping). As such, getters are one of the standard tools in the vacuum scientist's kit. The getter technology is quite old, the first patent and applications may be found at the end of the 19th century. Getters were widely used in the second half of the 20th century for UHV applications, mainly in the form of sputter-ion and titanium sublimation pumps.

In the case of a vacuum system, the trapping of gas molecules helps to maintain and improve the vacuum after the evacuation of atmospheric air. All metals, with only very few exceptions, react with atmospheric oxygen, but only a few (i.e., the getters) react with all the gases usually present in a UHV system; these include hydrogen, nitrogen, water vapor, carbon monoxide, and carbon dioxide. Noble gases and methane are not chemically reactive and therefore cannot be pumped by getters. Pumping of gas molecules results in a progressive increase of the getter surface coverage and, consequently, in a decrease of its pumping capacity. In the case of titanium sublimation pumps, this phenomenon is overcome by a periodic deposition of a fresh titanium film, sublimated from a hot filament. An alternative strategy consists in diffusing the adsorbed gas molecules from the getter surface into its bulk by heating; in this case the getters are referred to as Non Evaporable Getters (NEGs). The temperature required for this operation of regeneration (activation)

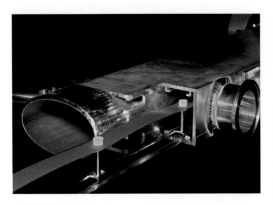

32. The LEP chamber consists of a cylindrical tube through which the beam passes, and the parallel chamber housing the NEG coated strip.

depends on the permeability of the getter to the different gases, an additional requirement that further limits the number of the usable materials for this technology. The large majority of NEGs are made of titanium and zirconium based alloys.

NEG pumping for LEP

NEGs were used for the first time to provide the main pumping of an accelerator for the Large Electron Positron Collider (LEP) at CERN, built in the 1980s in the same tunnel that today hosts the LHC. A total of 23 kilometers of metal strip, in sections 6-m long and 30-mm wide, coated on both faces with a zirconium-aluminium powder as the getter, were inserted along the entire length of the LEP vacuum chambers, in a separate volume but connected to the main beam pipe. The heating required for activation of the NEG was provided by circulating 100 A through the strip, which was suspended via electrical insulators. The cross section of the LEP chamber with the NEG pump is shown in Figure 32. The vacuum performance of LEP was excellent and very reliable. This solution was since adopted for other machines, in particular HERA in Germany.

Further developments of the NEG technology.

The solution adopted for LEP, however, is not free from inconvenience, the main one being the large size of the vacuum chamber required to host the NEG pump. Furthermore, the NEG required complicated mounts and was subject to the risk of electrical short circuits. The LHC improved on both these design limitations by replacing the NEG pump with a NEG thin film, coated on the internal surfaces of the vacuum chamber. This approach provides additional advantages, namely a larger pumping area and the elimination of the need to degas the chamber. Furthermore, a clean getter film would also decrease the degassing of the surface induced by the circulating beams, created when bombarding the surfaces under vacuum by electrons and ions.

The successful design of Ti-sublimation pumps is clear evidence that thin getter films may provide pumping when deposited under vacuum. But we did not know if a titanium thin film, if exposed to the air, could recover its chemical reactivity by heating under vacuum. For niobium films, however, we had some evidence that this is indeed the case. During the construction of LEP, a better production technique for the superconducting accelerating radiofrequency cavities was developed. During this work, copper cavities coated with a thin film of Nb were proposed and studied as an alternative to the more expensive "traditional" cavities made of bulk Nb. In the end, Nb-coated copper cavities were adopted for the energy upgrade, the second phase of the LEP project. Niobium is a good getter but becomes contaminated if deposited under poor vacuum conditions, thus leading to the deterioration of the superconducting performance.

A similar deterioration was also experienced by heating a coated cavity under poor vacuum, indicating the existence of post-coating gettering action, good news in view of the LHC design.

From the chemical point of view, the possibility of regenerating the chemical reactivity of a surface by heating depends on the permeability of the getter material for the gases to be pumped, which may be enhanced by increasing the temperature. Heating a coating implies heating the coated chamber; therefore only NEG films with an activation temperature compatible with the mechanical stability of the chamber may be considered. This requirement reduces the choice of the candidates for coating use, in particular for chambers made of aluminium. Because of its low density, aluminum is an interesting candidate for the construction of the vacuum chambers located at the intersecting points of storage rings, but unfortunately aluminum chambers cannot be baked at temperatures higher than 200°C. At the very beginning of the NEG coatings study many coatings were obtained with activation temperatures in the range from 250°C to 350°C; but it took much longer to find out that a titanium-zirconium-vanadium (TiZrV) film could be activated at 180°C, which was then adopted for LHC.

Production and performance
The coating is applied by atomic sputtering. Wires of the three elements are twisted together to form a straight rod, which is inserted in the middle of the chamber to be coated (Fig. 33). Atmospheric air is then evacuated and the chamber is filled with a rare gas (argon or krypton) at a pressure of about 1 Pa (10^{-5} atm). By applying a negative electric bias of a few hundred volts to the rod, a gas plasma is produced inside the chamber. Under the effect of the electric field, the getter (cathode) is bombarded by the positive ions, and atoms sputter off and coat the surrounding chamber.

After coating, the chamber is vented to air, the cathode is extracted and the chamber is ready for installation. The NEG coating will later be activated

33. The NEG coating facility for the LHC vacuum chambers.

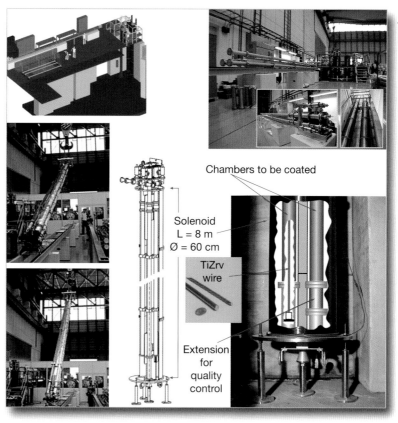

Chambers to be coated

Solenoid
L = 8 m
Ø = 60 cm

TiZrv
wire

Extension
for
quality
control

during the bake-out of the chamber, which is part of the standard commissioning process.

The NEG coatings performed according to our expectations. Inside the coated chambers, pressures down to the 10^{-12} Pa (10^{-17} atm) range were measured. Chambers exposed to synchrotron radiation at ESRF-Grenoble (France) showed an impressive reduction of radiation-induced degassing when compared to uncoated chambers, an effect which resulted in the much-improved performance of the associated experimental lines. Besides these anticipated benefits, an additional important bonus of NEG coatings is the very low secondary electron yield they display after activation (i.e., the number of electrons created when the material is hit by a beam particle), with values between 1.1 and 1.2, compared with values of about 2.0 for untreated vacuum chambers. Such a low value prevents the multiplication of electrons in the presence of circulating beams, which otherwise contribute to the detrimental electron clouds. For these reasons the decision was taken to coat all the room temperature chambers of the LHC with NEG films.

LHC and other applications
More than 1200 chambers, representing a total length of some 6 km of the LHC vacuum system, were coated with 1-μm-thick films. At the time when the coated chambers were needed, industry was not equipped for the job. To face the tight schedule, a dedicated facility was built at CERN. It consists of three cylindrical magnetron sputtering systems that allow vacuum chambers, with a maximum length of 7.5 m and maximum diameter of 60 cm, to be coated. Each unit consists of a vacuum pumping system, a manifold, a base support and a vertical solenoid (Fig. 33). The chambers were assembled on a special cradle, able to support up to four chambers if their diameter is less than 100 mm.

A 16-m-long assembling bench allowed the horizontal insertion of the cathodes in the chambers. The whole structure could then be lifted up with a crane and inserted into the solenoid, which is installed inside a 6.3-m-deep pit. In the simplest geometrical configuration, a three-wire cathode is inserted in the chamber and aligned along the pipe main axis. For more complicated structures, more than one cathode is necessary to guarantee uniform film thickness.

34. Drawing of one dipole-dipole beam vacuum interconnect.

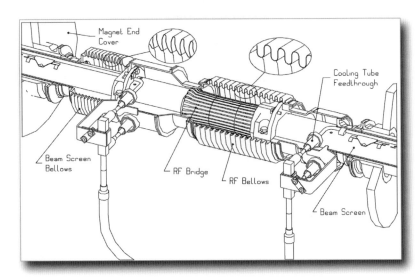

An average production rate of about 20 chambers per week was achieved. Quality of the production was monitored by coating sample surfaces simultaneously with the main chambers and analyzing them for thickness and pumping performance.

The work to develop the NEG coating technology is one of the outstanding success stories of the LHC, even before it entered into full operation. The coating technology is protected by patents, which have been licensed to five institutes and companies. These NEG coated chambers are now commercially available and frequently used for new accelerators. Other applications are envisaged in the fields of semiconductors and solar energy.

The cold interconnects

The cold arcs of the LHC consist of twin aperture dipole, quadrupole and corrector magnets in cryostats, operating at 1.9 K, as described in Chapter 4.1. The beam-vacuum chambers, which are embedded in the superconducting magnets, along with all connecting elements, require flexible "interconnects" between adjacent cryostats to allow for thermal and mechanical offsets foreseen during machine operation and alignment. In addition, the electrical continuity of the beam screens described earlier in this chapter must also be ensured by providing low impedance in the interconnect zones; this is done with the so-called RF bridges.

The design had to take into account (a) offsets expected during normal operation of the machine; (b) transient offsets during machine cool-down and warm-up; and (c) "exceptional" values due to possible faults, for example in the cryogenic system. The layout of a dipole-dipole beam vacuum interconnect is shown in Figure 34 with the main features labeled. Each interconnect zone contains two such beam lines in parallel.

Mechanical and electrical requirements
The mechanical requirements are mainly driven by the limited space of the 500 mm available between magnets to accommodate the thermal expansion and contraction, together with lateral misalignment of the magnets. The stroke of the interconnect module between room temperature and 20 K, showing the contracted and expanded bellows, is illustrated in Figure 35.

From the electrical point of view, two constraints had to be satisfied. The first one is to have a low longitudinal resistance (less than 100 $\mu\Omega$ at working temperature) between two consecutive beam screens. The second one is to shield the outer bellows, the convolution of which would otherwise be seen as an RF cavity by the proton bunches of the beam, hence inducing losses.

The solution
The mechanical aspects are covered by the use of two bellows: an RF bellows around the RF bridge which allows for the thermal contraction (up to 62 mm across one interconnection) and lateral offset (up to ± 4 mm) of the magnet cold masses; and a beam-screen bellows which absorbs differen-

35. Drawings of the contracted and expanded RF bellows, corresponding to the position at room temperature and when the LHC is operating at 20 K.

tial thermal expansion (34-mm nominal, but 75 mm in exceptional cases) between beam screen and magnet cold mass during thermal transients, preventing damage to the cooling tube feedthroughs. The minimized space in the interconnection zone imposes strong limitations on the length of both bellows units. The RF bellows is a hydroformed U-type profile. The size and axial stiffness were optimized for mechanical stability (buckling) and fatigue (thermomechanical cycles). For the beam screen bellows a very compact "nested bellows" was selected to compensate for the large transient axial offsets that occur during cool-down and warm-up. On the beam screen side, the cooling tubes are routed out of the vacuum chamber in a way that avoids all direct welding between the beam vacuum system and the helium lines. The end of the beam screen is equipped with a conical clamp to maintain it in place, and is finished with 3 μm of gold over the last 3 mm to optimize the electrical contact between the beam screen and the interconnection module. The various components that insure all the required functionalities are shown in Figure 34; Figure 36 shows the two extremities of a magnet, completed and ready to be interconnected.

The RF bridge itself consists of a set of copper-beryllium fingers sliding on a copper insert. Using copper-beryllium for the fingers allows them to act as springs and to exercise enough contact forces over the whole expansion range of the RF bridge. The ends of the fingers are coated with 5 μm of gold to optimize the electrical contact and avoid fusion due to induced currents during magnet quenches. The copper insert is coated with 3 μm of rhodium to avoid cold welding under vacuum. The RF bridge design concept was developed to avoid the high shear stress in contact fingers associated with the

37. The components of a plug-in-module (PIM).

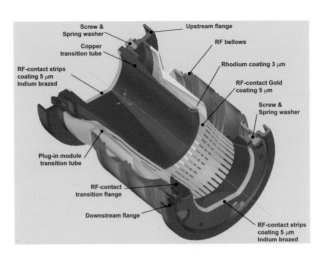

38. A mock-up of a plug-in-module in its nominal position.

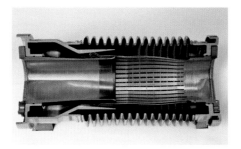

large lateral offsets required to re-align the machine when warm. The solution adopted was to provide clearance, rather than contact, when warm and to use a change in the chamber wall diameter to apply the contact finger force only when the magnets are cold. This can also be seen in Figure 37.

The RF bridge and the RF bellows have been assembled as an independent module, called a plug-in module or PIM, which could be assembled, cleaned and tested individually before installation. Figure 37 shows all components of a plug-in-module, Figure 4.38 shows a mock-up of a PIM in its nominal (extended) position at cold.

The plug-in-modules were delivered in a "compressed" state to make their installation between two magnets relatively easy. Figure 39 shows a compressed PIM during installation and the same after the compressing tooling has been removed. The sequence of installation in the tunnel has been optimized to cope with the limited space available in the interconnect zone. The final welding which makes the beam pipe leak-tight was done with automatic orbital welding machines, specially adapted to the limited space (Fig. 40).

39. The installation of a plug-in-module. Compressed as supplied to facilitate installation (left); and in place (right).

40. The orbital welding machine installed and ready to weld.

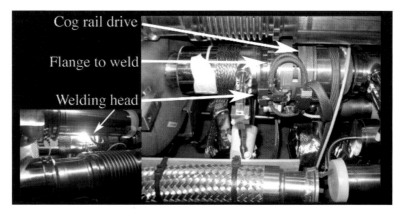

Difficulties encountered and overcome

The mechanical design of the PIMs had been extensively tested under all possible situations of cooldown and warm-up before the construction was launched. Despite all these precautions, a number of them failed during the first warm-up of LHC sector 7-8 in August 2007. A variable number of contact fingers buckled towards the inside of the beam pipe on 6 out of 366 modules, hence obstructing the beam path almost completely as can be seen on Figure 41.

A crisis task force was set up to investigate this rather catastrophic and completely unexpected behavior of the PIMs. All parameters were once again verified using sophisticated finite element calculations, which demonstrated that a PIM built as it was designed should not fail. Measurements were made, at cold, of the friction coefficients, which also showed values within the expected range. An analysis of the displacements due to thermal effects revealed that indeed some PIMs were more vulnerable than others, but still within design tolerances. One interconnect was equipped with remote measurement of the displacement during cool-down, which confirmed the calculated values. The position of all magnets was measured in the tunnel, and even if this survey campaign revealed a systematic displacement of the quadrupoles by 2 mm, this could still not explain the failure of the modules. The effort then continued with the verification that the modules have been built according to the specification, which was in theory the case from the manufacturing documents. It was found that an initial manufacturing error in the curvature of the end of the contact fingers was corrected by introducing a too large an angle in another part of the finger, thus the stability of the fingers could no longer be guaranteed under all circumstances. Figure 42 shows two pieces that highlight the difference between the specified and measured values for the erroneous curvature.

Finite element calculations, consolidated by a number of laboratory tests, using the "as-built" geometry of the fingers, confirmed the origin of the problem. The team developed a method to restore the geometry of the contact fingers for the modules not yet installed, but more than 80% of all modules were already welded in place when the problem was discovered. X-ray imaging was used to localize other faulty modules, but the shielding by the outside bellows of the interconnection reduces the accuracy of this method. For the longer term, X-ray tomography may become a valid method, as shown in Figure 4.43 on a laboratory test, but in the short term a perfectly dependable method was needed to find PIM failure during LHC warm-up. We came upon a very clever idea, consisting of "blowing" a small ball equipped with an RF transmitter through the entire arc and using the beam position monitors to follow the ball. The ball is "pushed" by connecting a mechanical pump to one end of the arc under test, thus creating a "vacuum" of about 200 mbar, enough to create a draft of 6 m s^{-1} which makes the ball roll at a speed of about 2 m s^{-1}. This method was successfully tested when sector 4-5 of LHC had to be warmed-up to connect the final focusing magnets. Twelve faulty modules were found, all in the most vulnerable locations, also confirming the expected statistics. Two more sectors sere tested recently, in which the quadrupoles had been previously re-aligned. Three faulty modules were found in each sector. This lower number confirms again that the calculations are correct, but also that the problem will persist until all vulnerable PIMs have been repaired.

Table 2. Summary of design criteria for the cold interconnect.

Design options	Reason
Sliding RF fingers	Shield the beam from the convolutions of the external bellows which are seen as RF cavities, while allowing thermal expansion during cool-down
Copper beryllium RF fingers	Provide a constant contact force over the full expansion of the module during cool-down
Copper insert	Minimize electrical resistance between two adjacent beam screens
Concept of "plug-in" module	Allow for easy installation in a tight space
Gold coating of the extremities of the RF fingers	Minimize electrical resistance between RF fingers and copper insert
Rhodium coating of copper insert	Avoid "cold-welding" between RF fingers and copper insert under vacuum

Prepared for the next steps

In this chapter, we summarized the requirements for the vacuum system and the main stages of the vacuum technology development for the LHC, from the use of standard techniques to the development of new technologies. The spirit of ingenuity of the team in charge is highlighted by the three special cases, where a very complex parameter field had to be considered to design the equipment and to implement improvements. These efforts have already led to the licensing of new technologies and to approaches that will find ever-new applications in particle accelerators and other scientific fields. Finally, we have seen that despite considerable effort to study, plan and implement a technological design, the unforeseen failure of a component is always possible. Thanks to the commitment and ingenuity of the team, we have been able to keep the impact of these failures to a minimum and to continue development, manufacturing and installation according to a very ambitious schedule. The next step will be to verify the quality of the vacuum system when submitted to the dynamic effects induced by the beams.

42. Sample contact fingers showing the critical parameters as specified and "as-built" for the PIMs.

43. X-ray tomography on a complete mock-up of an interconnection zone, showing its potential utility in detecting defective PIMs.

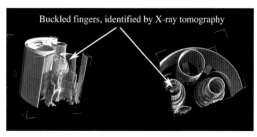

Buckled fingers, identified by X-ray tomography

4.3

THE TECHNOLOGY OF THE LHC

The Cryogenics Challenge of the LHC

Philippe Lebrun and Laurent Tavian

In order for the LHC to operate as planned, a major challenge would need to be overcome: the development of adequate cryogenic technology. As we have seen in previous chapters, the refrigeration system must operate at low enough temperatures to push the operating range of the Nb-Ti superconductors to new extremes, thus allowing the magnetic field strength to be extended beyond 8 tesla (Section 4.1). The LHC is a massive machine, with a cold mass weighing tens of thousand of tons, that needs to be cooled down close to absolute zero in order to benefit from all the physical properties of the superfluid helium. Nothing on this scale had ever been attempted; before entering into the details of the cryogenics, let us consider some of the implications of the refrigeration requirements:

- Approximately 80 tons of superfluid helium must be maintained at 1.9 K during the entire period of operation.
- The mass cooled by the helium to this temperature is 37,000 tons.
- Just pre-cooling this mass uses 10,000 tons of liquid nitrogen;
- The surface area cooled exceeds 40,000 m^2;
- All leaks of non-viscous superfluid helium had to be eliminated after tens of thousands of welds;
- A new family of efficient, low-temperature refrigerators had to be developed;
- Efficiency that allows a 0.1 K gradient over 3.3 km had to be achieved!
- The high-precision temperature sensors and instrument actuators have to function in a hostile radiative environment.

In this chapter, we will explore the science and technology behind these challenges, organized according to the aspects of magnet cooling, heat-load management, refrigeration, instrumentation, and fluids.

Superfluid helium from the laboratory to the field

A brief background on superfluid helium is presented in Chapter 1; here we recall some aspects of this history relevant to the technologies discussed later in the chapter.

Although the superfluid phase of helium was certainly produced by H. Kammelingh Onnes in his Leyden laboratory as early as 1908, when he lowered the pressure on the saturated liquid and unsuccessfully tried to solidify it, it took another twenty years for his disciple W. H. Keesom to establish the phase diagram of helium and identify the second-order phase transition, occurring at 2.17 K, from the standard liquid ("He I") to a second liquid phase

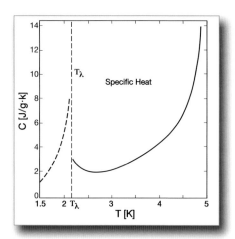

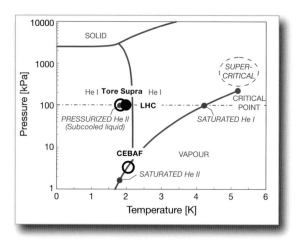

44. Specific heat of liquid helium, showing the lambda transition.

45. Helium phase diagram, showing working points of Tore Supra, CEBAF and LHC.

("He II"). A reason for this delay is that efforts were distracted by the unexpected discovery of superconductivity in pure mercury in 1911, soon followed by tin and lead. Thereafter, the laboratory studied materials at low temperatures in addition to their work on the thermodynamic properties of fluids, its original *raison d'être* in the wake of the pioneering work by J. D. Van der Waals, Kammerlingh Onnes' mentor [1]. There is no latent heat associated with the phase transition from He I to He II, but there is a peak in the specific heat, the shape of which led Keesom to name it the lambda point (Fig. 44).

In the 1930s, liquefaction of helium had diffused from Leyden to several other laboratories in the world, as interest in low-temperature research expanded. Thus in 1938, P. L. Kapitsa in Moscow, and J. F. Allen and A. D. Misener in Cambridge independently discovered superfluidity in He II, manifesting itself through totally unusual characteristics such as the absence of viscosity, thus enabling the liquid to flow through very fine-pore plugs (called *superleaks*); a very high thermal conductivity prevents the saturated liquid from nucleating bubbles in order to vaporize; and – even more surprising – the liquid is seen to creep upwards and above the brim of a container. In addition, they discovered the thermomechanical effect, i.e., the direct conversion of a thermal gradient across a superleak into a pressure difference, allowing helium fountains to form. As we see below, these properties have been put to good use in the cryogenic technologies employed for the LHC.

We note for the interested reader that this wealth of experimental observations proved a challenge for theoreticians to explain. F. London very early related superfluidity to a quantum effect and the superfluid phase to a Bose-Einstein condensate. L. Tisza proposed a phenomenological model of two interpenetrating fluids ("two-fluid model") which successfully accounted for the observed behavior, but had no microscopic foundation. L. D. Landau addressed the question of elementary excitations in the fluid, introducing the concept of quasiparticles which proved fecund in condensed-matter physics [2]. Only after the Second World War did these different approaches harmonize with the theory of quantum liquids, the explanation of the excitation spectrum, and the observation of quantized vortices in the superfluid.

With the development of superconducting magnets for scientific applications in the 1970s, it is quite natural that the lower temperature range and unique transport properties of superfluid helium be considered for refrigeration. Saturated superfluid helium, however, shows several technological shortcomings. First, it only exists under low pressure, thus requiring large-capacity vacuum pumps to maintain the pressure on the liquid bath, and absolute tightness to atmosphere in order to prevent inward leaks of air and moisture which would instantaneously freeze and block piping. Second, the low-pressure helium vapor exhibits low dielectric strength, a major drawback in electrical devices that operate at moderate or high voltage. Third, stories of superleaks frightened engineers and technicians who had to design and build helium-tight vessels using the standard construction techniques of cryogenic engineering, such as tungsten inert-gas welding of stainless steel. Still, the potential benefits of helium-II cooling of superconducting devices for reaching higher fields were a strong enough incentive for G. Claudet and his team in CEA Grenoble, France, to study, establish and promote its use as a technical refrigerant. To circumvent the question of sub-atmospheric pressure in the device cryostat, they proposed using pressurized *superfluid*, i.e., liquid at atmospheric pressure, sub-cooled by heat exchange with a heat sink consisting of a saturated liquid bath. The first sizeable project using this technique was the high-field hybrid magnet at the Service National des Champs Intenses in Grenoble in 1980. This constituted the working model for the first large-scale application of superfluid helium to the cooling of magnets, the Tore Supra tokamak which began operation in Cadarache, France in 1988 [3]. Tore Supra contains 520 kg of superfluid helium, cooled by a 300-W-at-1.8-K refrigerator.

In parallel with the cooling of high-field magnets, the emergence of high-field superconducting radio-frequency (RF) cavities for linear accelerators raised the question of limiting alternating-current losses in the superconducting wall of the cavity (see Chap. 4.5). When subject to rapidly varying current, superconductors, in spite of their zero direct-current resistance, exhibit power dissipation. Since this dissipation grows exponentially with temperature in the range of interest, operating the cavity at lower temperature immersed in a saturated helium-II bath proves to be globally efficient, in spite of the higher thermodynamic cost of refrigeration. This solution was retained for the re-circulating linear accelerators of CEBAF, built in 1995 at the Thomas Jefferson Laboratory in Newport News, USA. In contrast to Tore Supra, housed in a single hall with the refrigeration plant adjacent, CEBAF features the extended geometry of an accelerator, spreading over several hundred meters. It contains 12 tons of superfluid helium, cooled by a 4.8-kW-at-2.0-K refrigerator. Both Tore Supra and CEBAF established industrial feasibility of the superfluid helium technology, and paved the way for its implementation on a novel and much grander scale at the LHC.

Magnet cooling

The main reason for superfluid helium cooling of the LHC magnets is the lower temperature, which extends the operating range of the Nb-Ti superconductor and thus results in field strength beyond 8 tesla (Chap. 4.1). However, the rapid drop in specific heat of the conductor at low temperature also requires making use of the transport properties of superfluid helium for thermal stabilization, heat extraction from the magnet windings, and heat transport to the heat sink. With its low viscosity, superfluid helium can permeate the windings, where it buffers thermal transients thanks to its high specific heat (2000 times that of the conductor per unit volume). The excellent thermal conductivity of the fluid (peaking at 1.9 K with typically 1000 times that of good copper) enables it to conduct heat without mass transport, i.e. with no need for fluid circulation or pumps. In order to benefit from these unique properties,

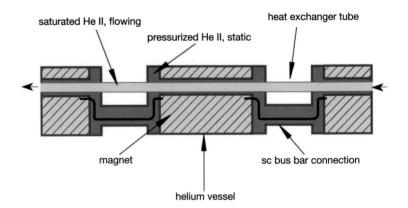

though, the electrical insulation of the conductor must have sufficient poros-
ity providing percolation paths, while preserving its main functions of mechanical resistance and dielectric strength. These conflicting requirements are met
by a multilayer wrap of polyimide film with partial overlap (explained in more
detail in Chap. 4.1).

The practical thermal conductivity of superfluid helium remains finite,
and therefore insufficient to transport refrigeration along a full LHC sector,
the 3.3-km stretch of tunnel cooled by one refrigerator. Moreover, we know,
since the time of J. Fourier, that heat can only be transported across a thermal
gradient, so that the temperature at the refrigerator must be lower than the
1.9-K upper bound of the warmest magnet, located some 3.3 km away. In
order to limit the thermodynamic penalty of having to produce refrigeration at
too low a temperature, the total temperature gradient used for the cooling of
a sector is a mere 0.1 K, thus requiring an extremely efficient cooling scheme.

The LHC magnets operate in static baths of pressurized superfluid helium
close to atmospheric pressure, according to an unconventional cooling
scheme. This high-conductivity single-phase static liquid is continuously cooled
by heat exchange with saturated two-phase helium flowing in a heat exchanger
tube, made of cryogenic-grade copper threading its way over the length of the
magnet string (Fig. 46). The deposited heat is eventually absorbed quasi-
isothermally by the latent heat of vaporization of the helium flowing in the
tube. The residual temperature gradients are those generated by the thermal impedance at the tube wall (known as Kapitsa resistance) and by the pressure
drop in the two-phase flow. To minimize the pressure drop, the scheme is implemented in 107-m-long strings of magnets, parallel connected to supply and
recovery headers running along the tunnel in a separate cryogenic distribution
line, located along side the magnets, spanning the length of each sector
(Fig. 47). Other benefits of this cooling scheme are
- the little transverse space it occupies in the magnet cross-section;
- the capacity to absorb large heat loads such as generated by current ramps
  or resistive transitions; and
- the thermal decoupling of the magnets as soon as the flow dries out, thus
  limiting quench propagation in the string and protecting the neighboring
  magnets.

This unconventional cooling scheme was investigated and tested experimentally for flow stability and heat transfer performance, and finally validated
in a prototype magnet string. It is now routinely operating up to expectations
in the LHC tunnel (Fig. 48).

Thermal insulation and heat-load management

With their cold mass of 37,000 tons at 1.9 K, presenting an external surface area of some 40,000 m² (five soccer fields!), the LHC magnets must be efficiently insulated to meet the refrigeration requirements: this is the task of the cryostats (Fig. 49). Conventional thermal design and cryogenic construction techniques usually applied in clean workshops on laboratory-size equipment had to be implemented here on an industrial scale, involving field assembly underground in sometimes difficultly controlled conditions. Priority was therefore given to robustness of design and reliability of construction over sheer thermal performance.

As in other low-temperature apparatus, the first concern was to reduce gas conduction, which could represent a major path of heat in-leak. This is achieved by placing the cold mass inside a vacuum enclosure and maintaining it under high vacuum (a residual pressure below 10⁻⁹ atm). Thanks to the low temperature, the high vacuum is maintained by a phenomenon referred to as cryopumping, in which all gases other than helium are removed by condensation on the cold surfaces; this works provided the helium circuits are tight, the latter condition being ensured in the long run by integrally welded stainless steel construction.

48. Temperature profile along a 3.3 km sector: all magnets are isothermal within 0.05 K.

49. Artist view of the LHC main dipole magnet in its cryostat.

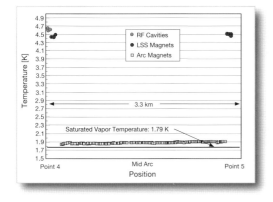

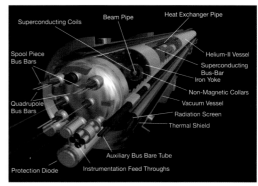

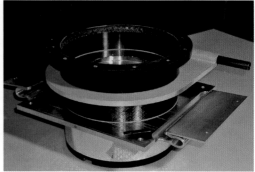

50. Assembly of LHC magnets in their cryostats.

51. Glass-fiber/epoxy composite support post for LHC cryomagnets: thermal intercepts at 50-70 K and 5 K reduce the conductive heat in-leak to the top plate at 1.9 K.

52. Interconnection between two cryomagnets with vacuum sleeve open and thermal insulation removed, showing welded seams and compensator bellows on piping.

Thermal radiation is reduced by interposing a radiation shield, cooled at 50 to 70 K by circulation of gaseous helium. The power radiated from the room temperature vacuum vessel to the shield and the residual heat in-leak from the shield to the cold mass are further limited by multilayer insulation blankets, prefabricated in industry and installed during cryostat assembly (Fig. 50).

The cold mass is supported on column-type support posts, made of glass-fiber epoxy composite working in compression, with heat interception at both 50-70 K and 5 K temperatures in order to limit the solid conduction at 1.9 K (Fig. 51).

In order to accommodate the cold mass shrinkage by 0.3% during cool-down (i.e. 45 mm on a 15-m-long magnet, or 8.4 m on the cold length of each sector), the support system incorporates low-friction sliding surfaces, and the interconnecting pipe-work features compensator bellows, which must also take alignment tolerances during installation (Fig. 52). The assembly of the cryo-magnets in the tunnel to form a continuous string incurred the proper execution of 65,000 low-resistance splices on the superconductors, and of 40,000 helium-tight welds on cryogenic piping, carried out by an industrial contractor. This was achieved by proper preparation of methods and tooling; maximum automation; training of personnel; establishment and enforcement of quality assurance procedures; and strict control. Overall, the measured linear heat in-leak to the cold mass of the LHC cryomagnets is as low as 0.2 W/m, in agreement with the calculated value. This is a remarkable achievement.

Besides the leaking in of heat by the above processes, the LHC magnets are also subject to dynamic thermal loads induced by a number of processes:

- residual dissipation in the conductor splices;
- eddy currents and hysteresis during charge and discharge;
- circulating-beams loads such as synchrotron radiation, beam image currents, loss of particles from the beam halo, and bombardment by electrons resonantly accelerated by the beam potential (the so-called electron cloud).

Since some of these losses dominate at high beam energy and intensity, the magnet apertures are equipped with beam screens (Chap. 4.2) cooled by circulation of supercritical helium between 5 and 20 K, to intercept most of them at a higher temperature than the 1.9 K of the cold mass, thus reducing the associated thermodynamic penalty. Primarily installed for cryogenic reasons, the beam screen is also used as the protective baffle of the distributed cryopump constituted by the wall of the beam pipe at 1.9 K, so as to improve substantially the dynamic vacuum seen by the circulating beams.

Large-capacity refrigeration

In this section, we discuss the helium refrigerators that cool the LHC sectors. This section is more technical than the others of this chapter, and because of space limitations we assume that the reader has some background in thermodynamics.

A common practice for comparing different cryogenic systems is to transform their refrigeration duties in equivalent entropic capacity at 4.5 K by applying Carnot factors. For the LHC, this amounts to a total of 144 kW at 4.5 K, thus making it the world's most powerful helium refrigeration system. In practice, a sector refrigeration plant is composed of a 4.5 K refrigerator with an equivalent capacity of 18 kW at 4.5 K, coupled to a 1.8 K refrigeration unit producing a cooling capacity of 2.4 kW at 1.8 K. Four out of the eight 4.5 K refrigerators are recovered from the previous LEP accelerator, and suitably adapted to LHC duties. The helium refrigerators of the LHC operate in mixed duty in order to absorb a variety heat loads at different temperatures (cooling of the magnets, beam screens, radiation shields and current leads).

Previous experience at CERN with large 4.5 K refrigerators, delivered by European industry in the framework of turn-key contracts, has demonstrated their dependable performance, good efficiency and high operational reliability. Consequently in 1997 CERN issued a functional and interface specification for the procurement of four such refrigerators. The adjudication rule took into consideration, besides capital investment, the integrated costs of operation over a period of ten years, thus giving a premium to efficiency, and, indirectly, compactness. The LHC 4.5 K refrigerators show efficiencies around 29% with respect to the Carnot cycle, corresponding to a coefficient-of-performance around 230 (i.e., 230 W electrical consumption per W refrigeration at 4.5 K). Based on a modified Claude cycle with three pressure levels (typically 0.1 MPa, 0.4 MPa and 2 MPa), a refrigerator consists of a compressor station and a cold box (Fig. 53). Each compressor station has five to eight oil-lubricated screw compressors, water refrigerants for helium and oil, as well as a final oil removal system achieving a residual content of a fraction of a part per million. The installed electrical input power is about 4.5 MW per refrigerator. The vacuum-insulated cold box houses aluminium plate-fin heat exchangers and eight to ten turbo-expanders to provide the cooling capacity. To prevent any pollution from air which could enter the cycle and become solid at cryogenic temperature, switchable 80-K adsorbers remove up to 50 parts per million of air and a 20-K adsorber removes remaining traces of hydrogen and neon. Switchable dryers are connected at the ambient temperature inlet of the cold box to remove humidity.

A novel requirement set by the LHC project is the efficient and reliable production of 1.8-K refrigeration in the multi-kW range. This can only be achieved in practice through combined cycles making use of sub-atmospheric cryogenic compressors and heat exchangers. As early as the mid 1990s, CERN stimulated industry to develop low-pressure heat exchangers and hydrodynamic cryogenic compressors of different designs, through procurement of several prototypes. In particular, the cryogenic compressors involve advanced mechanics and hydrodynamics, with impeller wheels rotating at up to 50,000 rpm. At

53. 4.5-K-helium refrigerator for LHC: compressor station (top); and cold boxes (middle and bottom).

their low temperature of operation, they cannot be lubricated and to avoid any contact, their rotating assembly operates suspended on active magnetic bearings. The prototypes of low pressure heat exchangers and cryogenic compressors were thoroughly tested in nominal and off-design modes. Design and optimisation studies were also performed in liaison with industry on refrigeration cycles matched to the expected performance of full-size machinery.

This preparatory work permitted the launch of the procurement for eight 1.8-K refrigeration units in 1998 (Fig. 54). The overall coefficient-of-performance of these 1.8-K refrigeration units, once attached to the main 4.5 K refrigerators of the LHC, reaches around 900 (i.e. 900 W electrical consumption per W refrigeration at 1.8 K), corresponding to efficiencies around 18% with respect to the Carnot cycle. The cycles are based on 3 or 4 stages of axial-centrifugal cryogenic compressors (Fig. 55) raising ambient temperature through heat exchangers, before its final compression to above atmospheric by volumetric screw compressors, thus limiting the size of the latter. The cryogenic compressors are equipped with active magnetic bearings operating at ambient temperature, with rotational speeds from 12,000 rpm to 48,000 rpm for the warmest stages. The machines now operate routinely in the accelerator.

Cryogenic instrumentation

The cryogenic operation and monitoring of the LHC requires a large number of sensors, electronic conditioning units and actuators. About 15,000 instrumen-

54. 1.8 K refrigeration unit for LHC: (top) room-temperature compressors, (bottom) cold box with cryogenic compressors.

tation channels are located inside the machine tunnel and must therefore withstand the radiation environment, imposing strict radiation qualification procedures. While a large variety of sensors and actuators are commercially available, specific tests on prototypes were performed to select the suitable types. Furthermore all tunnel electronics are custom designed to be radiation tolerant.

The very limited temperature excursions allowed along the cryomagnet strings require the implementation of precision cryogenic thermometry on an industrial scale, with several thousand channels exhibiting long-term robustness and reliability. The overall measurement uncertainty must be less than ± 10 mK with respect to the operating temperature of 1.9 K. Following the dedicated construction and commissioning of cryogenic calibration facilities of metrological class at CERN and CNRS Orsay, France, several types of sensors were tested for performance in LHC environmental conditions on statistically significant ensembles. In particular, the effects of neutron irradiation at cryogenic temperature and of thermal cycling were investigated on several hundred thermometers, in order to select appropriate solutions for the project (Fig. 56). The stringent requirements on temperature measurement, once applied to signal conditioning, cannot be met by commercially available equipment. Several conditioner architectures able to simultaneously provide the large dynamic range, high accuracy, stability and tolerance to radiation levels were investigated, prototyped and series produced in application-specific inte-

111

The Technology of the LHC

56. Cernox™ temperature sensor (right) with thermalisation plates (left) and support blocks for pipe mounting (center).

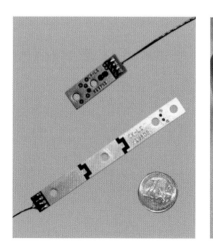

112

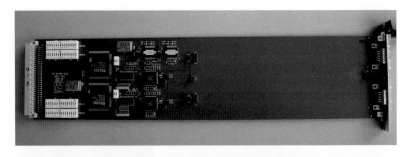

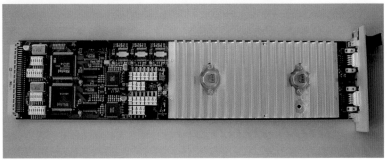

57. Radiation-tolerant conditioner for temperature signals (top) and level signals (bottom), using submicron ASIC technology.

grated circuit (ASIC) versions (Fig. 57). The complete system is now commissioned and delivers the expected performance.

Cryogenic fluids

Helium is mainly contained in the magnet cold mass, where a minimum amount of superfluid is necessary for enthalpy buffering, and in the cryogenic distribution line running in parallel to the cryomagnets. The total helium inventory of the machine is about 130 tons, with about 60% in the superfluid state. The complete storage of all the helium would have too expensive, requiring a large facility. To limit investment and reduce the environmental impact, it was decided to provide physical storage for only half of the inventory. In total, 90 medium-pressure gaseous helium vessels totalling 1500 m^3 capacity (Fig. 58) and two liquid helium reservoirs totalling 240 m^3 capacity (Fig. 59) are available for storage, distributed over the different technical sites around the LHC. The gaseous helium pressure vessels are interconnected by a medium-pressure line, allowing inventory to be shuffled around. During machine shutdown when all the sectors must be emptied, loss of helium is economically not acceptable. Consequently, the helium which cannot be stored on-site is recovered (and later re-supplied) by gas companies in the framework of ad hoc supply contracts featuring "virtual storage" clauses.

Liquid nitrogen is only used for pre-cooling the accelerator down to about 80 K, and for regeneration of the adsorbers. This avoids permanent liquid nitrogen consumption and the need for corresponding delivery logistics. For reasons of safety, liquid nitrogen is not distributed in the LHC tunnel, but vaporized by heat exchange with gaseous helium at ground level. For cooldown of a 3.3 km sector, the 600 kW pre-cooler uses 1,260 tons of liquid nitrogen, vaporized at rates up to 8 tons per hour. Simultaneous pre-cooling of the complete machine requires about 50 liquid nitrogen semi-trailers per day, i.e. about one truck delivery every half-hour, day and night. The liquid nitrogen delivery logistics is therefore essential for a continuous cool-down operation. During yearly maintenance shutdowns, however, the sectors of the LHC not requiring interventions stay idle and remain cold, thus alleviating the liquid nitrogen requirement for re-cooling.

Conclusions

In order to build the LHC, it was necessary to conceive, design and build the most advanced low-temperature refrigeration system in the world. Although the basic technology and thermodynamics behind the conception of the installation were well known, the successful construction of the refrigeration stages brought significant technical progress in the field, with some major technological extrapolation from previous experiments and installations. The result is a feat of modern technology on a scale that would have been unimaginable only a few years ago, which will benefit other large cryogenic projects such as the ITER tokamak, as well as the FAIR and XFEL superconducting accelerators.

References and further reading
[1] H. Kammelingh Onnes, *Investigations into the properties of substances at low temperature, which have led, amongst other things, to the preparation of liquid helium*, Nobel lecture (1913).
[2] S. Balibar, "The discovery of superfluidity", *Journal of Low-Temperature Physics* **146** (2007) 441-470.

[3] G. Claudet & R. Aymar, "Tore Supra and helium II cooling of large high-field magnets", *Advances in Cryogenic Engineering* **35A** (1990) 55-67.
[4] Ph. Lebrun, "Cryogenics for the Large Hadron Collider", *IEEE Transactions on Applied Superconductivity* **10** (2000) 1500-1506.

58. 250 m³ storage vessels for gaseous helium at medium pressure (2 MPa).

59. 120 m³ vacuum-insulated storage reservoirs for liquid helium.

4.4

THE TECHNOLOGY OF THE LHC

Moving the Beam into and out of the LHC

Volker Mertens and Brennan Goddard

In this chapter, the reader will learn about the proton transfer and injection into the LHC, and about the ultimate fate of the high-energy particles, i.e., the "dumping" of the 7 TeV proton packets at the end of the experimental cycle. The story of what happens between these two points in time is told in other chapters. Our description is divided in three basic parts. The first considers the transfer of the 450 GeV beam from the SPS to the LHC injection facility; then, the injection system itself is described in some detail, stressing the mechanical and technical challenges that the team faced; and final part describes the elegant concept and implementation of the remarkably complex system required for the safe disposal of the high-energy proton beams.

Injecting protons into the LHC

As mentioned in Chapter 1, the protons are first accelerated to 450 GeV in the Super Proton Synchrotron (SPS); their subsequent transfer to the LHC is a complicated three-dimensional problem. With the relative positions of the SPS and the LHC, it is necessary for the beams to travel many kilometers through beamlines in new tunnels before reaching their destination, where they have to be precisely injected into the LHC ring. The challenges include:

- precisely transferring the beam and placing it onto *a nominal* orbit inside the LHC ring, taking all possible sources of errors into account;
- maintaining the small beam *emittance* – meaning that the particles are confined to a small volume and are characterized by the same energy and direction – by an accurate matching of the optics at the injection point;
- designing flexible transfer optics in order to accommodate future LHC design modifications;
- avoiding beam contact with LHC elements which could lead to damage or a quench;
- designing, testing and constructing a 450 GeV injection system, including the injection kickers;
- reusing, for budgetary reasons, of as much existing material as possible from the LEP accelerator.

The beam transfer lines

Studies on how to best transport the beam from the SPS to the LHC began in the mid 1980s. The former LEP beam-transfer tunnels TI 12 and TI 18, originally built for electrons and positrons with an energy of 20 GeV, could not be

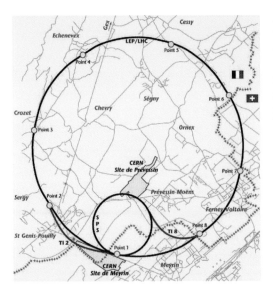

60. Situation of the LHC injection transfer lines TI 2 and TI 8 (red lines; sketch).

re-used because the maximum bending radius of the 450 GeV beam was too large to fit their geometry. Various transfer-line configurations were investigated in the early design phase, one of them even implying a polarity reversal of the SPS between the filling of the two LHC rings. The use of cryogenic magnets in the transfer lines was also considered, but, in the end, room-temperature magnets were selected for reasons of lower operational complexity and overall greater economy, since the transfer lines have to operate only during the short periods of LHC filling, about twice daily.

The final configuration meant the construction of two new beam transfer lines, TI 2 and TI 8 (Fig. 60). These are the longest beam transfer lines ever built, with a combined length of 5.6 km, 80% of the circumference of the SPS itself. A significant fraction of beam line components and other equipment was recovered from earlier, closed-down installations.

Geometrical layout and design constraints
The source protons for the clockwise beam of the LHC ring leave the SPS through a fast extraction system before entering the new TI 2 tunnel. About 2.6 km later it joins the LHC tunnel some 200 m upstream of the ALICE experiment at Point 2. One design requirement for TI 2 was to pass below the end of CERN's above-ground lab "Site de Meyrin", where the access shaft PMI2 was built. This hole, some 60-m deep and 18-m wide, was not only used to excavate the TI 2 tunnel, but also to lower most of the LHC machine components, including the 15-m-long main dipoles (seen earlier in Fig. 23a).

During the test borings along the future path of TI 2 an underground river bed was found below today's "Le Lion" river, now covered with several tens of meters of rubble ("moraine") left after the last Ice Age. To avoid civil engineering complications and long-term problems with water ingress, a Z-shaped vertical profile (Fig. 61) was adopted for this tunnel to remain in the more solid limestone ("molasse"). This is why TI 2 has significant vertical bending, even though the SPS extraction and the LHC injection at point 2 are nearly at the same height above sea-level.

The beam delivered to the counter-clockwise LHC ring is extracted from the SPS through a transfer line used in common with the primary proton beam line to the CNGS (CERN neutrinos to Gran Sasso) target, used to produce

61. Vertical profile of TI 2 (left) and TI 8 (right; in green the CNGS tunnel).

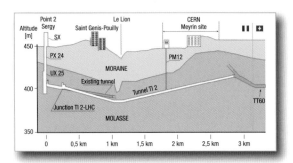

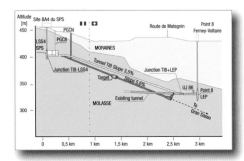

neutrinos which are directed towards the Gran Sasso Laboratory in Italy. After the branching point the TI 8 line continues over 2.4 km to the LHC injection point near the LHCb experiment at Point 8. The TI 8 tunnel is also entirely lodged in the limestone. Most of it has a slope of 3.5%, as the injection at Point 8 is some 70 m lower in altitude than the SPS, due to the inclination of the LHC.

Civil engineering

In total, the beam transfer to LHC required the excavation of more than 5 km of new tunnels and enlargements, involving 60,000 m^3 of material, and the use of 21,000 m^3 of concrete. Excavation for TI 8 began in autumn 1998 with a civil engineering shaft near the SPS. The first enlarged part of the tunnel and some adjacent underground works were excavated using machines known as "road headers", while for drilling through the 2.3 km towards the LHC, a tunnel-boring machine was used. Excavation finished in June 2000 and was followed by lining with concrete, leaving a finished tunnel 3 meters in diameter. By contrast, TI 2 was entirely excavated by road headers. More details about the civil engineering are found in Chapter 3.

Transfer line elements

To build TI 2 and TI 8, over 700 magnets were required. The 348 dipoles forming the main horizontal bends, the 178 main quadrupoles and the 98 corrector magnets were all built by the Budker Institute for Nuclear Physics at Novosibirsk (BINP), in the framework of the contribution of the Russian Federation to the LHC, and transported by lorry over 6,000 km to CERN. The main quadrupoles are 1.4 m long and have an inner diameter of 32 mm. One of the design criteria for the 6.3-m-long main dipoles was to be able to re-use the main high-current LEP power supplies. At a nominal current of 5340 A and a gap height of 25 mm they reach a magnetic field of 1.81 tesla. All the upper coils of the main arc are connected in series; thus, the current flows from one coil to the next, thus avoiding massive individual cables. The dipoles are interconnected by short copper braids, while the quadrupoles are bridged by water-cooled bus bars.

The various magnets are supplied with power from either the SPS end or the LHC end, to reduce cable length and cost, and to best use the space available in the surface buildings to house the voluminous power converters. Most of the converters (and even some of the power cables)

117

were recovered from LEP, overhauled and fitted with modern electronics. With the exception of the small correctors, all magnets are pulsed together with the SPS cycle. The total reactive power of TI 2 and TI 8 is about 30 MVA (mega-volt-amperes). In addition to the main magnets, the bulk of the vacuum system is also provided by BINP, while the ion pumps have been recovered from LEP. Every second quadrupole is fitted with a pump, which provides for a beam vacuum of about 10^{-10} atm.

Because of the small emittance of the beam, and the fact that it only makes a single pass, the mechanical apertures of the lines could be relatively small – sometimes no bigger than a postage stamp. These small apertures and the length of the lines require careful control of the trajectory and appropriate beam instrumentation, in particular near the injection region.

On average, every second quadrupole is equipped with beam position monitors and correctors, one each for the horizontal and vertical planes. A number of beam screens, thin sheets made of alumina (Al_2O_3; for low-intensity beam) or titanium (for higher intensities), with observation by a CCD camera, allow the transverse profile and size of the beam to be measured, from which the beam emittance and local optics values can be calculated. The data from the screens also complement the position monitoring. At regular intervals along the lines and at strategic points, ionization chambers detect beam losses. Beam current transformers at the beginning and the end of each line measure the flux of particles.

Various protection devices are placed along the lines, like beam dumps, safety stoppers and collimators. The beam dumps consist of graphite cylinders in a copper housing, surrounded by 21-ton cast-iron shielding. They are used firstly as safety equipment – in conjunction with other safety measures – to block the beam during the stage when the transfer lines or the LHC is not ready to be accessed. In addition, they serve to set up precise SPS extractions and transfer-line positioning before the beam is permitted to enter the LHC. To protect the elements at the end of the lines and in the LHC against damage from a potentially poorly steered beam, each line includes a system of six collimators with 1.2-m-long graphite jaws closed to gaps of only a few millimeters.

An important amount of infrastructure and ancillary hardware components, like lighting, general electricity distribution, water cooling, ventilation, emergency cut-off, evacuation alarms, telephone, control cabling, etc., completes the transfer line equipment.

65. Installation of the first magnet in the enlarged transfer tunnel TT40 on 19 February 2003. Günther Kouba (green helmet) explains to Lyn Evans the functioning of the new installation convoys.

Installation

From the outset we aimed to set up and commission the LHC beam-transfer equipment in stages. The first installation began on 19 February 2003 with the first magnet put onto its supports, just downstream of the new SPS extraction. Some 100 meters of tunnel were already fully equipped and commissioned before the SPS closed for operation, and this stretch was successfully commissioned with beam on 8 September 2003.

Maneuvering space in the transfer lines was tight, with a variety of installation situations and magnet types. To cope with this, a completely new transport system was conceived, based on compact "buggies" with a maximum payload of 9 tons each, driven by in-wheel motors able to turn on the spot. Up to 4 buggies can be coupled together and equipped with additional accessories like a

66. Installed section of TI 8 consisting of a main quadrupole (blue), followed by a corrector (green) and a series of main dipoles (dark red), all built by the Budker Institute for Nuclear Physics (BINP) in Novosibirsk, Russia (photo V. Mertens).

small crane. An auxiliary vehicle accompanying the buggy train carries the control and power electronics, the backup and buffer batteries and the driver seats. All vehicles can be optically guided by following a line on the tunnel floor, a prerequisite to move such a convoy along the narrow transfer tunnels with reasonable speed (about 3.5 km/h). Once the convoy has arrived in front of the magnet position the buggies move laterally under the beam line in between the supports feet and the magnet is put onto or lifted from its jacks using air cushions mounted on top of the buggies. To ease the maneuver in the confined space the whole convoy can also be remotely controlled from hand-held panels. Due to its versatility this system has also been used extensively for the CNGS neutrino-line installation, as well as for many of the warm magnets in the LHC.

Installation of the magnet system in TI 8 was started in December 2003 and finished in May 2004, followed by installation of the other line elements and performing the thousands of power, water and vacuum connections.

First beam test

We were fortunate to have been able to carry out an early beam test of the new long line, in particular as the SPS was scheduled to take a long break until Spring 2006 to free resources for the installation of the LHC. This served, firstly, as large-scale test-bed for the hardware components and the control concepts to be used later on in the LHC; secondly, it allowed us to gain experience about the behavior of the line.

To minimize the impact on the ongoing LHC installation work, the beam tests were scheduled and spread over two weekends, on 23/24 October and 14/15 November 2004. The excitement and relief in the control room was great when, after the final preparation and checks, on 23 October 2004 at 13:39, a single bunch of 5×10^9 protons travelled down the whole line up to the beam dump a few meters before the LHC tunnel, on the first attempt, with-

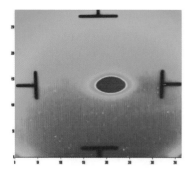

67. Computer image of the first beam, which arrived on 23 October 2004, at 12:15, at the last screen monitor in TI 8, a few meters away from the LHC tunnel.

68. Director General Robert Aymar and the LHC project leader, Lyn Evans, joined some of the people involved in the completion and testing of the LHC beam transfer lines in the old SPS/LEP control room, during the first beam tests of TI 8 on 23 October 2004.

out the need for any steering (Figs. 67 and 68). During the measurement program that followed, the basic optics model of the line was well confirmed. The trajectory stability was found to be very good and the layout of the beam diagnostics, which performed well, was shown to be appropriate. The new control system, with its extensive array of applications, performed excellently and greatly facilitated the tests.

Installation of the upstream part of TI 2, up to PMI2, was performed between the end of 2004 and late Spring 2005, followed by a preliminary hardware commissioning phase. The downstream part had to remain empty of line elements to ease the transport of LHC magnets. Installation resumed in April 2007 and was finished in early August 2007. After final preparation the first beam passed down TI 2 on 28 October 2007, again without any need for specific steering.

The injection systems

In both injections, the beam approaches the LHC from outside and below the machine plane (Fig. 69). The beam is deflected by a last series of dipoles, already located in the LHC tunnel, towards a series of five septum magnets (MSI) which deflect the beam horizontally by 12 mrad under the outer ring. A series of four fast kicker magnets (MKI) then deflects the beam vertically by 0.85 mrad onto the orbit.

The injection area is fitted with several different items of dedicated equipment:

- In order to set up the injection with low-intensity beam and to protect the LHC against the high-intensity nominal beam in case of the malfunctioning of the injection kickers, an injection beam stopper (TDI) is placed 15 meters upstream of the superconducting recombination dipole D1, supplemented by an additional shielding element.
- The protection against injection errors is further complemented by two collimators, near the quadrupole Q6 on the other side of the interaction point.
- To allow precision steering in this aperture-critical area, and to position the beam optimally inside the main elements and to control their correct functioning, the injection area is fitted by appropriate beam instrumentation equipment, such as beam screens (before and after the septa and the kickers and in front of the TDI), beam-loss detectors, and beam-position monitors.

Injection septa

The septum magnets are designed to deflect the beam horizontally, as mentioned above. The magnets chosen for this function are normal-conducting Lambertson septa, where the iron of the magnet yoke separates the zero-field

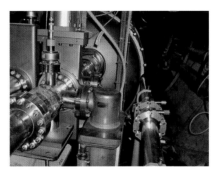

69. Near its end, TI 8 (to the right) passes very close to the cryostat of the quadrupole Q6, which was slightly modified (photo V. Mertens).

region from the high-field region. Each magnet is about 4.5 m long, and a total of 5 are required per beam. The full series of 12 magnets (including spares) together with the 33 septa required for the dump system (described in the next Section) were completed and transported overland from Russia by truck in an epic journey, and delivered safe and sound to CERN between 2002 and 2003 at the rate of about one per week.

Another challenge was the construction of the vacuum chambers that carry the circulating beam through two small 64-mm-diameter holes in the iron of the septum yoke. The chambers had to be built to very precise tolerances and then baked in-situ to $250\,°C$ in order to reach the vacuum levels needed. The chamber had to shield the circulating beam from stray magnetic field in the septum hole. The chambers were built from mu-metal, which has the correct magnetic properties but is difficult to handle mechanically. This material had to be rolled into shape, carefully welded into cylinders, coated with a thick layer of copper to reduce the impedance seen by the beam and finally coated again with the NEG material used for vacuum purposes on all warm sections of the LHC (Chap. 4.3). A special thin heating system developed at CERN and a thermal insulation layer were then added to the outside of the tube, and the chamber then had to be inserted into the septum for the final assembly and welding of the end flange. At all stages of the process, the highest quality standards had to be maintained, and it was a great relief to the vacuum group after 50 of these delicate objects had been successfully installed into the injection and beam dump septa.

Injection kickers
After passing the septa and the superconducting quadrupole Q5, the beam arrives at the injection kickers, consisting of four fast-pulsed magnets per injection, whose role is to vertically deflect the beam to its final orbit. The construction of the injection kickers was one of the major challenges of the injection system, with many of the components specially developed for the purpose. A long prototyping phase was required. The synchronization, generation and control of the duration of the magnetic pulses represented a technological breakthrough – almost perfect rectangular pulses with almost no ripple are required in order for the kickers to carry out their function satisfactorily. The magnets are each housed in separate vacuum tanks, through which both beam pipes pass. The high-voltage pulse generators and part of the power and control electronics are located in the adjacent underground gallery.

To give the reader an idea of the time scales involved, during its operation, the LHC will be filled with 12 batches of either 5.84 μs or 7.86 μs duration, to be deposited successively on the machine circumference, with 11

70. Sequence of four injection kickers at Point 8 (photo V. Mertens).

gaps of 0.94 μs between the batches, during which the kicker field has to rise to its nominal value. To limit the loss of beam coherence (emittance blow-up) at injection, reflections and ripple of the field pulse must stay below $\pm$ 0.5%, which constitutes a very stringent requirement. One final gap of 3.0 μs in the LHC beam distribution leaves time to reduce the kicker field back to zero; it also provides the time needed for the rise time of the beam-dump kickers.

Each kicker magnet is about 3 meters in length and consists of a series of 33 cells (Fig. 70). The cells are designed to achieve the characteristic impedance within the space constraint imposed by the 540-mm-diameter tanks; each cell consists of two ceramic plate capacitors, 210 mm in diameter, with a contoured rim, fitted between high-voltage and ground plates, electro-polished to provide good voltage holding capabilities. The plates are spaced by ceramic-metal insulators. Figure 71 shows one of the cells during assembly.

The selection of materials for these components was critical. The viability of using ceramic capacitors in ultra high vacuum had been demonstrated with out-gassing tests. Good vacuum quality of some 10^{-13} atm during operation was ensured by bake-out at 300 °C and vacuum firing or heat treatment in air of most of the magnet components. To permit baking, the magnet and current conductors have to be made of stainless steel, because aluminum is unsuitable for such temperature cycles. Low inductance damping resistors are connected in parallel with the cells; these were specially developed for UHV compatibility and consist of two counter-wound Kanthal® wires on a ceramic rod. The ferrite cores are made from high permeability and high resistivity NiZn, with magnetic properties specially adapted to this application.

In order to reduce beam impedance while allowing the fast field rise, the injected beam passes through a 3-m-long extruded ceramic pipe with resistive wires lodged in grooves in the 4-mm-thick pipe wall (Fig. 72). The grooves are produced directly during the extrusion process. The wires provide a path for the image current and thus screen the ferrite yoke against beam induced heating. They are directly connected to the standard vacuum chambers of the machine at one end and via a decoupling capacitor of 300 pF at the other, using the ceramic pipe itself as the dielectric.

71. Assembly of a high voltage cell
(photo N.Garrel).

72. Technicians adjusting RF contacts
at the end of the ceramic beam tube,
before closing the vacuum tank
(photo V. Mertens).

The kickers are fired by thyratrons of three-gap type, chosen for minimum size and easy maintenance, installed in independent tanks with isolating transformers for heaters, reservoirs and grid biasing. Because thyratron switches are gas tubes, they may self-fire ("erratic pulses") or skip a cycle ("missing pulses"). In these cases the beam to be injected receives the wrong deflection angle, and is then absorbed by the injection beam stopper, described below.

Injection beam stopper

The injection beam stopper (TDI) is normally used to set up the injection timing and steering with a low-intensity beam, but must also be able to protect the LHC in case of malfunctioning of the injection kickers. It is located about 70 meters downstream of the injection kickers. The active part of the TDI consists of two 4.2-m-long absorber jaws, the position of which can be accurately adjusted around the circulating beam, such that particles with small injection errors will be caught. Two additional collimators, positioned near Q6 on the other side of the insertion, extend the protection of the TDI.

Each TDI jaw consists of a sequence of segments of absorbing materials mounted in an aluminum frame. Eighteen boron nitride blocks (hBN) with a combined length of 2.9 m are followed by 600 mm of aluminium and 700 mm of copper. Heat deposited by the beam is conducted to the tank by flexible copper strands. To reduce beam impedance, the hBN blocks are coated with 3 μm of titanium. Due to the proximity to the ALICE and LHCb experiments, the background from residual gas scattering in the TDI must be kept low, requiring a special design for the vacuum and bakeout systems. In case the full beam is absorbed by the TDI, the maximum temperature rise is 640°C in the hBN, 150°C in the aluminium block and 190°C in the copper. Extensive simulations showed that this construction could absorb full-intensity batches without causing damage, and in most cases without causing a quench of the downstream superconducting magnets.

Once past the TDI jaws, the protons are placed on their nominal orbit within the LHC. So far we have seen how, after the proton packets are provided with initial energy in the SPS, these are then transferred by tunnel to the injection system, which accomplishes the delicate task of adequately placing the protons into the LHC ring. Later in this book, in Chapter 4.5, we will discuss how these proton packets are accelerated to their final experimental energy; but to complete the story of the proton transfer into and out of the LHC, we now turn our attention to a critical aspect of the LHC, namely, its beam-dump system.

The LHC Beam Dump system

Removing the beam from the LHC is a task both delicate and forceful. The beam must be steered with great precision out of the accelerator, but the deflection of the proton beam at the LHC energies requires very strong pulsed magnetic fields.

The LHC stored-beam energy presents a major challenge in the design of the accelerator. At 7 TeV/c with nominal intensity, the stored energy in each of the LHC beams is 350 MJ, equivalent to the kinetic energy of a TGV train travelling at 100 km/h, or the chemical energy contained in 35 kg of Birchermüsli. This is over one hundred times higher than the beam energy stored in other high-energy proton accelerators such as the SPS, HERA and the TEVATRON (Fig. 73).

The beam revolution time is less than 0.0001 second, so that the instantaneous power when the beam is stopped is 3.9×10^{12} Watt, 50 times higher than the combined output of all the power stations in the United Kingdom. The power density is also enormous, since the transverse beam dimensions are sub-millimeter. The enormous power density would destroy any solid material object placed in its path, and the accidental loss of even a small fraction of a percent of the LHC beam has the potential to damage machine components.

The LHC relies on the correct functioning of interdependent systems like magnets, cryogenics, beam control, vacuum, collimators, power supplies, experiments and many other elements. In case of problems with any of the many thousands of critical LHC sub-systems, the potentially destructive beam might be lost within a small number of turns, on timescales down to a few milliseconds. The LHC does possess a collimation system designed to safely absorb the continual low intensity flow of particles lost at the periphery of the circulating beam, but this would be extensively damaged were the full beam to be lost on a timescale of seconds. A reliable way is therefore needed to quickly remove the circulating beam from the accelerator and to dispose of it safely. For this purpose the LHC has a dedicated beam dump system in Point 6 of the machine that can extract the beams, dilute their power density to a safe level and deposit them on special absorber blocks. All this has to be performed within one beam revolution period of the ring with an extremely high reliability.

Many other accelerators make use of an internal beam dump, which requires only a fast pulsed kicker-magnet system to deflect the beam and a dump absorber block placed outside the machine aperture. This concept is straightforward, compact and robust but is simply not possible for the LHC. The energy deposition would be orders of magnitude too high for any solid material to resist, and it would be impossible to operate such an internal system with the efficiency necessary to avoid quenches of nearby superconducting magnets.

124

73. Stored beam energy for different proton accelerators.

74. Original sketch of the early design for an external LHC beam dump, 1984 (courtesy E. Weisse).

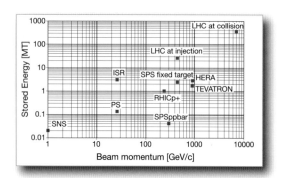

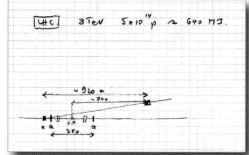

First thoughts on the dump

For the LHC it was therefore decided from the outset, in the early 1980s, that a system would be built to transport the beam to an external dump (Fig. 74). This would be placed far away from the LHC machine to provide enough increase in the beam size and allow for the use of a solid dump absorber block, and to be far enough from the superconducting magnets to minimize the risk of quenches during the operation of the beam dump.

Detailed studies of the LHC beam dump system started as early as 1983. In 1984 the options for injection, beam transfer and beam dumping aspects of the LHC and its injector chain were examined in more detail. The co-existence in the same straight section of the beam dump system with an LHC physics experiment was already excluded, and the difficulty of maintaining the crossing between inner and outer rings with the beam dump system was pointed out, with implications for the design of the rest of the ring.

The dump-system design evolved over the following decade, in collaboration with the designers of the SSC beam dump. The main design as late as 1993 still assumed crossing between inner and outer rings in the center of the dump insertion, and used strong superconducting separation dipoles that had the novelty of containing a field-free coil window through which the extracted beams would pass. When the crossing between rings in the dump insertion was abandoned, the dump design was simplified to a more conventional layout using normal-conducting magnets.

At the very early stage of the LHC design it was envisaged to have two straight sections dedicated to beam dumping, with the beam always being extracted from the outer ring. This was over-optimistic regarding the number of available straight sections and the design soon converged to both systems housed in a single straight section. This was initially foreseen to be around Point 3; however, this straight section is located under the Jura mountains with difficult conditions for civil engineering, not compatible with the extensive tunneling and excavations required for the beam dump system. A proposal to build a combined dump/collimation insertion in Point 3 was evaluated but dropped, and after further study in 1990, where Point 7 was also considered as a possible location, the LHC beam dump system settled in Point 6 as its home.

After all these iterations the concept and location of the LHC dump system was essentially fixed in 1995, and since then the dump system concept has changed only in small details.

Converging on a design

The actual LHC beam dump system comprises several different components (Fig. 75). Fast kicker magnets deflect the beam in a very short time in the hor-

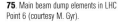

75. Main beam dump elements in LHC Point 6 (courtesy M. Gyr).

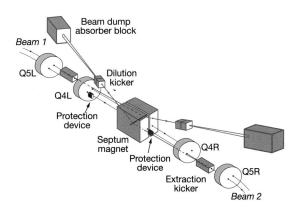

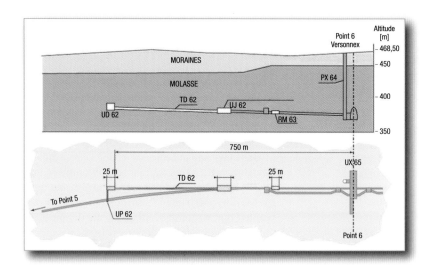

76. Layout of the new underground excavations for the beam 2 dump.

izontal direction by about 50 mm, enough to move it just outside of the circulating beam aperture and into the high-field aperture of a septum magnet, which remains permanently powered. This septum provides a strong vertical deflection, which after several hundred meters is enough to lift the beam above the superconducting magnets comprising the main LHC ring and into a dedicated extraction line.

During the passage of the extracted beam, additional pulsed kicker magnets are used to dilute the beam. The beam is then transferred through a large diameter, 700-meter-long evacuated vacuum line to increase the transverse beam size to approximately 1.5 mm and to spread the bunches further apart. Finally, the diluted beam impacts the dump block, where the energy can be absorbed without damage. A variety of special beam instrumentation systems allow the system to be set-up and operated with beam.

The LHC beam-dump system uses some leftover infrastructure from the LEP, with the kicker generators housed in the service galleries parallel to the main LHC tunnel. However, the drift distance required between the extraction kickers and the absorber block is of the order of a kilometer. In addition, the absorber block needs to be surrounded by a considerable volume of shielding for radiological reasons, and it must be located away from the main LHC tunnel. For these reasons, two new 3-m-diameter transfer tunnels (Fig. 76), each 325 metres long, had to be constructed tangential to the existing tunnel in Point 6. The new tunnels terminate in the dump caverns UD62 and UD68, which are each 25 metres long and 9 metres in diameter. The caverns are equipped with overhead cranes to allow remote handling of the beam dump absorber and the shielding blocks.

To ensure that all particles can be extracted from the LHC without losses, the succession of beam bunches contains a particle-free abort gap with a length of 3 millionths of a second. The extraction kicker is triggered such that the field increases from zero to the full value during this gap in the beam. This needs an accurate synchronisation between the LHC radiofrequency system (described in Chapter 4.5) and the beam-dump kickers, to make sure that the beam is deflected with exactly the required angle. Since the kicker can only be triggered at the start of this gap, the delay between the request for beam dumping arriving in Point 6 and the beam extraction can be up to two LHC turns.

The heart of the beam dump

The extraction kicker magnets and their generators are the heart of the beam dump system; a photo is shown in Figure 77. Building the pulsed kicker magnets and their power supplies involved a number of cutting-edge technologies which were the result of long collaboration between CERN and high technology industry world-wide.

The 15 magnets which kick the beam into the septum have to be very reliable and robust, and strong enough to deflect the rigid 7 TeV proton beam. The magnets are built from thin steel sheets, wound tightly into cylinders which are impregnated with resin for mechanical stability and then cut to the precise shape needed for the magnetic field. The magnet is air-insulated, and assembled around a 1.6-meter-long ceramic chamber. The chambers were produced in Japan and carefully shipped to CERN for internal coating with a thin layer of titanium which serves to reduce the impedance seen by the circulating beam.

With the single-turn coil, a current of about 18,500 A is needed to reach the nominal magnetic field of 0.34 T. This would be difficult even for a normal magnet, where the current is slowly increased over seconds to the required value. However, here the technical challenge is multiplied by the fact that the field has to increase from zero to its full value in less than 3 millionths of a second, in close synchronism with the other magnets. The magnet current then has to stay at the required value for at least one turn of the accelerator, while all the particles are extracted.

Although the LHC beam dump kicker magnets have been designed with a very low inductance, 30,000 volts are still needed to reach the nominal current. This voltage needs to be switched on in less than a millionth of a second after a request to perform a beam dump is received. This places a heavy burden on the switches used to turn on the magnets, and the switch technology is one of the key parts of the beam dump system. The switches need to hold-off tens of thousands of volts for many hours. Then, when they are energised, they have to immediately start conducting tens of thousands of amperes in less than a millionth of a second. In addition, for the LHC, it is very important that the switches do not spontaneously trigger, since this would provoke an unwanted beam dump which would not be synchronised with the abort gap.

Designing performing switches

To meet these requirements the switches had to be developed exclusively for the LHC. The solution chosen was to use a stack of solid-state switches built of silicon wafers. Each wafer was specially treated to reduce the turn-on time, and is able to conduct the tens of thousands of amps needed. However, since a single wafer can only hold-off about 3,500 volts, a stack of ten wafers arranged in series (but triggered in parallel) is needed to make up a single

The Technology of the LHC

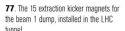

77. The 15 extraction kicker magnets for the beam 1 dump, installed in the LHC tunnel.

78. CERN engineer M. Mayer measures one of the ceramic vacuum chambers for the beam dump extraction kickers.

79. A dilution kicker magnet being inserted into its highly-polished vacuum tank, previously used in LEP.

80. The passage of the beam-dump vacuum tube between the Q5 magnet and its cryogenic and powering feedthroughs.

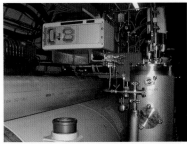

81. The beam 2 vacuum tube for the dumped beam passes above the main LHC ring and into its own tunnel.

switch. A magnet is energised by two switches, each of which can conduct the full current in case the other fails.

With the beam deflected out of the LHC aperture, the next job of the dump system is to extract it from the plane of the LHC ring, lifting it clear of the bulky superconducting magnets forming the LHC arc. This is done with *Lambertson septa*, of very similar design to those used for the injection system (described earlier in this Section) and built under the same contract, with a total of 15 required per beam. The septa for the two LHC beams had to be installed alongside each other in a back-to-back configuration because of the limited space between the beams of only 194 mm.

After the beam is safely out of the LHC ring there is still one final active task to perform, since even with a drift of around 700 meters the beam would still be intense enough to damage the absorber block. To reduce the energy density the beam is diluted by sweeping it across the face of the absorber, in much the same way that an electron beam in a cathode-ray TV set is scanned across the screen to produce the image. In the case of the LHC the scanning is done by two sets of kicker magnets, using very similar design and technology to the extraction kickers, but providing sinusoidal pulses which combine to sweep the beam in an "e" shaped Lissajous figure (seen in Figure 82). The trace is about 100-cm long, which dilutes the peak energy density deposited in the dump absorber by a factor of 50. The diluter magnets are contained in the large vacuum tanks which originally housed the LEP high voltage separators.

Once past the dilution kickers the expanding beam is transported to the absorber inside a large diameter vacuum chamber. This pipe, built from industrial-standard tubing with a diameter up to 600 mm, has to pass through the LHC tunnel very close to the complicated cryogenic and power feedthoughs for the superconducting magnets (Fig. 80). Threading the line through this zone was a big job for the integration teams, who had to model every detail of the tunnel, the services and the installed equipment in three dimensions to make sure that everything could be accommodated. The size and weight of the vacuum chamber posed its own problems, especially for supporting and handling the objects.

The vacuum line ends in a 15-mm-thick entrance window made of carbon composite with a 0.2 mm steel backing sheet, thin enough to allow the diluted beam to pass without damage, but which can maintain the static one-bar pressure difference between the vacuum line and the dump block which is kept at atmospheric pressure of nitrogen. The design and fabrication of this unusual window was done at CERN, including the assembly and final welding of the 600-mm-diameter pieces.

The end of the line

For the absorber block the material choice is paramount. Beam impact in material produces particle cascades due to nuclear and electromagnetic interactions, and the energy deposition of a particle distribution is a function of material and geometry. This can be calculated with computer programs, and the temperature increase in the material can be estimated.

The choice of material for the dump absorber was a critical one for the project, and was one of the main system design choices from the beginning. Alternatives which were evaluated over the course of the project included lithium, for its low density and large enthalpy; H_2O either in the form of liquid water or as a large block of ice; and also more exotic substances such as high-temperature ceramic NbC or beryllium. The final choice of carbon (in the form of graphite) as absorber material was made on thermo-mechanical criteria, cost and practical grounds. The low density of carbon means that the nuclear and electromagnetic particle cascades spread out over a longer distance in the beam direction, reducing the peak energy deposition and resulting in lower temperatures. Carbon also has an extremely high melting point, in excess of 4000 K. Finally, composite carbon materials can be produced which exhibit excellent mechanical properties, with good thermal shock resistance.

82. Simulated sweep shape and temperature increase in the central part of the carbon beam dump absorber after full beam impact.

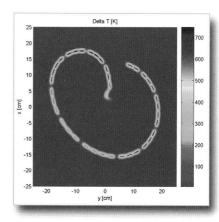

Under normal beam-dilution conditions at nominal beam intensity the maximum temperature increase in the block is limited to 750 °C. Each beam-dump core comprises a segmented carbon cylinder of 700-mm diameter and 7700-mm length, shrink-fitted in a stainless steel jacket with 12-mm-wall thickness (see Figure 4.83). The core assembly is connected upstream to the beam dump vac-

83. Left: Installation of one of the dump absorber blocks into the shielding. The front face of the graphite block is visible. Right: The configuration of the installed absorber blocks.

uum line by a quick-disconnect flange, while a leak-tight titanium window closes the downstream end.

Beam dump security

For radiological reasons the inner absorber needs to be surrounded by around 900 tons of shielding. An imaginative solution was found by reusing the old ISR magnets (from the Intersecting Storage Ring facility, decommissioned in 1984) which were filled with concrete and equipped with handling lugs. Each block weighs 24 tons and a total of 35 are arranged beneath, around and above each dump absorber block, leaving just the entrance free for the incoming vacuum chamber. It is appropriate that components from the ISR, the world's first hadron collider, are still being used in the LHC (Fig. 84).

The beam dump system also has to be protected against itself! Synchronisation errors between the extraction kick and the abort gap are not impossible, and another risk is that the extraction kickers might spontaneously trigger. Also some of the beam might find its way into the abort gap, for example due to losses during the RF capture process. In these cases, firing the dump kickers would risk damage to other components around the machine, as the beam is swept across the aperture. In particular the extraction septa would be hit by about 20 bunches if the beam were to be dumped asynchronously.

To protect against such scenarios, the beam-dump insertion is equipped with dedicated devices which intercept particles mis-steered by the extraction kickers. A six-meter-long fixed graphite and titanium absorber protects the extraction-septum magnets, sitting squarely in front of the exposed iron septum, between the circulating and extracted beam-vacuum chambers (Fig. 85). Another six-meter-long graphite absorber is installed in front of the first superconducting quadrupole, to intercept any particles that could leak into the arc of the LHC or beyond. This large object can be moved in and out of the beam to adjust the protection as a function of the beam energy, which is not constant during the acceleration phase. It is supplemented by a double-jawed carbon-composite collimator, which allows the beam position to be accurately constrained between the jaw faces. These mobile devices must be positioned with respect to the beam axis with an accuracy of better than ±0.3 mm at 7 TeV.

In the event of an unsynchronised abort, these two protection devices must safely absorb several tens of undiluted 7 TeV bunches, and must suffi-

84. LHC beam dump shielding blocks made from recuperated ISR dipole magnets filled with concrete.

ciently attenuate the shower of particles that constitute the secondary cascade, such that the downstream accelerator components are not damaged. They are subject to severe thermal loading, which provides difficult engineering challenges, in particular regarding the dynamic mechanical stresses. These elements have had to be designed on the limits of technical feasibility – simultaneously absorbing as much energy as possible to protect the elements beyond them, while limiting the energy deposition to a level below their own damage threshold (Fig. 86). The conceptual and mechanical designs have been the subject of extensive energy deposition and finite-element dynamic-stress simulations, requiring several iterations in the design.

Monitoring operations
The beam dump also has its own nervous system, which monitors the state and controls the different elements. This uses a combination of industrial and CERN-built controls, optimized to provide an extremely reliable way of

131

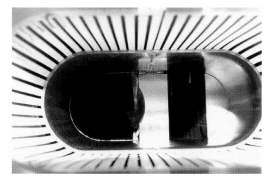

85. Beam-dump protection device installed in front of the extraction septum. The three-cm-wide gap for the extracted beam is seen at the right.

86. Mechanical thermally-induced stresses in horizontal (left), vertical (middle) and longitudinal (right) planes, vertical and longitudinal planes in a beam dump protection device, calculated 10 millionths of a seconds after impact of 20 LHC bunches.

87. The two beam-dump-trigger synchronization pulses locked to the incoming radio frequency signal for the first time, with a stability of better than 10 nanoseconds.

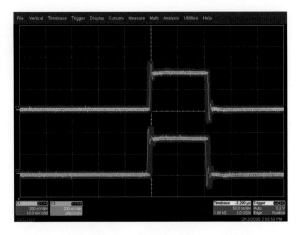

generating the triggers for the different magnets with the correct delays, and of making sure that the system parameters are all within the required tolerances (Fig. 87). The LHC energy changes over the course of its acceleration cycle but the geometry of the dump line is fixed. Therefore the active elements (extraction kickers, extraction septa and dilution kickers) must change in strength in proportion to the beam energy. A beam-energy tracking system takes the energy reference from two main LHC dipole circuits in the adjacent arcs, and distributes the reference settings to the dump elements. It also makes continuous checks of the measured settings of these elements against a different reference, since an error with the energy tracking could result in the full LHC beam being deflected at a totally wrong angle. Many other important functions are performed by the dump controls, including the synchronization of the trigger pulse with the abort gap and the distribution of all the trigger signals. The system also monitors the switches and the currents measured in the magnets, to make sure that all the magnets operated correctly. It also ensures that the system can be armed, which involves an intricate choreography with the beam interlock system and the LHC injection systems.

The complexity and criticality of the dump system mean that most of its functionality cannot be modified by the machine operators. The different settings are coded into the logic circuits which control the equipment, and only a small range of general functions can be executed from the control room, for instance to arm the system or to perform a beam dump at a specific time. A lot of development of high-level software was nevertheless needed, since there are many interfaces with other LHC systems. The dump must be able to fire in close synchronization with the injection of a beam into the LHC, and the web of connections to and from the other parts of the LHC had to be carefully defined, implemented in cable, fiber-optic or software, and then painstakingly tested during the LHC commissioning.

To set up the dump with beam and to monitor the performance of the components, the system also includes different types of beam instrumentation. Some of this is standard equipment replicated around the LHC ring, like the many hundreds of beam-loss monitors placed along the dump lines. Other instruments are more special – the luminescent screens placed along the extracted beam line are all individually tailored to their particular location, with different screen sizes and shapes. At the end of the dump line, just before the entrance window, sits a huge beam screen, 600 mm in diameter, which stays

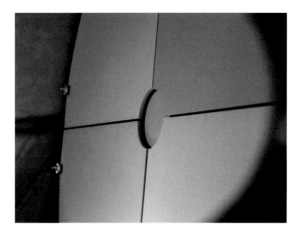

88. The 60-cm-diameter alumina screen at the end of the dump line, which will allow the trace of the extracted beam to be seen by the LHC operators.

permanently in place (Fig. 88). A camera pointed at the screen will allow the LHC operators to see the passage of the protons, and a few months after the startup the "e" shape of the dumped LHC beam will be a familiar sight on the computer consoles in the CERN Control Centre.

Success thanks to international collaboration

In this chapter, we have seen how the hadron beams get into the LHC from the SPS, and also how they are disposed of safely after the operational cycle has ended. By treating each of the transfer, injection and beam dump system elements in some detail, the authors hope to impress upon the reader the mix of physics, electrical engineering and materials science that has been required to reach our goals of providing high-quality packets of protons into a stable orbit inside the LHC, and of then being able to remove this huge stored beam energy in a controlled way.

A mention must be made of the many hundreds of people who have contributed to the design, construction and commissioning of the LHC transfer lines, injection systems and beam dump. They came from across CERN, from European industry and beyond, and from accelerator and research institutes worldwide, and provided a unique confluence of expertise in engineering disciplines as diverse as nuclear, high voltage, materials, thermo-mechanical, vacuum, beam dynamics, electro-magnetism, controls and software. Their marvelous contributions have ensured that these critical components of the LHC are now ready for the new challenge of operation with beam.

89. A young Eberhard Weisse, head of the CERN Beam Transfer group from 1981 until 2001, who designed the LHC beam dump system. Here he is at the controls of the SPS accelerator in the late 1970's.

The Technology of the LHC

References
[1] E. Weisse et al., "Design of the LHC beam dump," *Proc. 3rd European Particle Accelerator Conference*, Berlin, Germany (1992), p.1545.
[2] B. Goddard, "The Beam Dumping System," in Chap. 17 of Vol. 1 of the LHC Design Report *The LHC Main Ring* , O. Bruening et al., Eds., (2004) CERN-2004-003.

[3] R. Schmidt et al., "Protection of the CERN Large Hadron Collider," *New Journal of Physics* **97** (2008), NJP/230233/SPE.
[4] N.A. Tahir et al., "Impact of a 7 TeV/c Large Hadron Collider Proton Beam on a Copper Target," *Journal of Applied Physics* **97** (2005) 083532.

4.5

THE TECHNOLOGY OF THE LHC

Capturing, Accelerating and Holding the Beam

Trevor Linnecar

Particle beams, from the point of view of a physicist, must be as "bright" as possible with the maximum number of particles in the smallest volume, to maximize the particle collision rate and give the best chance of observing rare particles, possibly from new physics. The characteristics of the two counter-rotating beams – such as intensity, bunch spacing and size – are defined in the injector chain, and these properties must be maintained and controlled as far as possible in the LHC both during acceleration and during the long time "in store" at high energy. Proton beams, unlike beams of lighter electrons, are not strongly damped by synchrotron radiation and can become unstable when interacting with the surrounding equipment. In this Chapter, we discuss in some detail the radio-frequency (RF) technology that has been developed to accelerate and stabilize the beam. The RF systems installed at Point 4 are used to capture the beam from the moment of injection, accelerate the beam to top energy, confine it in store, and, perhaps most importantly, to control its stability and general characteristics.

Our ability to master these packets of high-energy protons is a triumph of modern engineering [1-3]. The technology described in this chapter can best be appreciated in the context of the orders of magnitude involved in the manipulation of the circulating protons:

- We use electric fields, oscillating at radio frequencies up to 400 MHz, to create stable volumes of space to confine individual packets of 10^{11} protons.
- In the beams there are 2808 packets of protons to manage when the LHC is operating, each separated by a distance of about 7.5 m.
- These protons are moving at a speed close to the speed of light.
- The position and stability of each packet are maintained to high precision by using finely controlled feedback loops.
- During acceleration the energy of each beam has to increase by 6.55×10^{12} eV, approximately 300 MJ, in 20 minutes, during which time the particles travel a distance further than the round trip to the sun.
- Detectors have been developed to precisely determine the position of each of these packets in space as they speed through the LHC.
- All the complex equipment must remain working faultlessly for many hours at a time.

The Technology of the LHC

Challenges

The basic concept is to use RF structures and their associated systems to provide the electric fields that accelerate the particles. This is a well-established art, and over the years many clever devices and techniques have been developed. The RF system of the LHC builds on this past but pushes the design beyond previous limits. Below, we summarize the main challenges in terms of the particle-beam dynamics:

• Beam at injection – the beam, which may be injected into the machine off-center and at the wrong energy, must quickly be brought under control in all three dimensions, before it loses its density.

• Beam stability – the very intense proton beam induces voltages both in the RF cavities themselves and in the vacuum chambers and other equipment. We have to prevent these induced voltages from acting back on the beam, thereby rendering it unstable.

• Beam loading – the voltage induced in the RF cavities by the intense beam can greatly exceed the accelerating voltage. As mentioned, this can lead to a loss of stability, and it also poses a risk of severe damage to the RF equipment if uncontrolled; a large amount of RF power is required to keep beam loading under control.

• Beam lifetime – mastery of beam size is very important to obtain the highest luminosity. If the particle bunch is too long, the beam density decreases and luminosity is lost. If it is too tight, then collisions between particles in the bunch (intra-beam scattering) will blow up each bunch transversely and again reduce luminosity. In addition, a sufficiently noise-free RF system is needed so that it does not excite the beam. Noise can then be artificially introduced to tailor the dimensions as required. Long lifetime also implies extremely high reliability. If only one cavity (out of the eight for each beam) becomes unavailable during store, for any reason, then the beam must be dumped.

These challenges, along with the basic acceleration and storage requirements, provide the parameters upon which the design of the cavity structure and RF power systems are based. As we will see, there is a broad range of feedback control systems, both slow and fast, that must be incorporated to ensure that it all works.

Acceleration and the principle of phase stability

Before discussing the technology in more detail, we will provide some of the basic concepts behind the use of radio frequency in particle acceleration (Fig. 90). The beam of charged particles is accelerated by an electric field. Radio frequency power, at a frequency F_{rf}, (e.g., 400 MHz for the case of the LHC) is fed into the RF cavity and creates an alternating electric field along the particle path. The cavity itself is basically a sphere with two openings to allow the beam to pass through. Particles that pass through when this alternating field is zero remain unaffected; otherwise they are accelerated or decelerated. If the RF frequency is an exact multiple, h, of the revolution frequency, then the particle always arrives at the same amplitude of electric field. Earlier particles will see a higher electric field which kicks them onto a longer orbit around the ring and makes them arrive later on the next revolution, while those that are later will see the opposite field and will come back earlier. As a result, a bunch of particles forms around the central point, with the particles moving around this point at the quite slow (e.g., 20 Hz) *synchrotron frequency*. The maximum distance that particles can be from the center and still be held in this bunch defines the bucket. Particles outside this bucket are lost. The maximum bucket

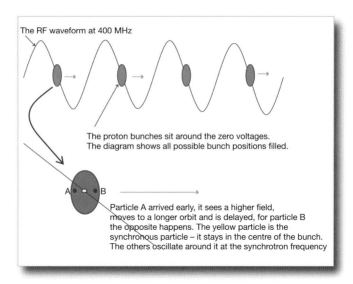

The RF waveform at 400 MHz

The proton bunches sit around the zero voltages.
The diagram shows all possible bunch positions filled.

A   B

Particle A arrived early, it sees a higher field,
moves to a longer orbit and is delayed, for particle B
the opposite happens. The yellow particle is the
synchronous particle – it stays in the centre of the bunch.
The others oscillate around it at the synchrotron frequency

length is $1/F_{rf}$. By displacing the RF field slightly in time with respect to the bunch, all particles receive on average a positive kick and are accelerated, the bucket length shrinking slightly in the process. Since the electric field repeats at the RF frequency, a maximum of h bunches can be placed around the ring.

RF Parameters

From the injection point, the beam is accelerated from the energy of 450 GeV to 7 Tev within the time of 20 minutes. Bearing in mind that, each second, the beam rotates about 11,245 times, the average increase in energy each time it passes the RF system is 485 keV. The 2808 bunches of 10^{11} protons will have acquired a total energy of approximately 340 MJ, and the power needed to carry this out will have been about 275 kW [4].

Just providing a high electric field (i.e. the accelerating voltage) in a copper cavity at room temperature would require significant power, far higher than the power required for the acceleration itself. For example, if we consider the typical value of an accelerating voltage of 2 MV, in a room-temperature cavity this would typically require about 250 kW, even in a cavity of optimum shape, with the power all being lost in the resistance of the copper in the cavity walls. At superconducting (SC) temperatures, this resistance is lowered enormously, and the lost power can be as low as 15 W! There are still considerable savings in power, even though the cryogenic plant liquefying the helium at the rate necessary to keep the cavity cold requires of the order of 100 kW for this operation. However, the big advantage of going to SC temperatures is the design of the cavity; because we no longer have to deal with the power deposited in the cavity, we can relax on the design parameters defining voltage in order to optimize the cavity in other ways.

A particular example of one of these design parameters is the stored energy. The SC temperatures of the cavities allow us to work at high stored energy, which is favorable for beam stability. It is true that, as the stored energy of the electric field in the cavity increases, more power is required for a given voltage; but, under these conditions, we have far more resistance to transients induced by the high-beam current. These transients come about because the beam consists of trains of bunches (high current) followed by gaps (no current). These voltage changes are extreme and can perturb the individual bunch positioning and the collision points of the experiments. Indeed, the control of these transients is very power demanding; this is why it is favorable to work at high stored energy: high energy minimizes both the relative transient and the power required to control it. High stored energy is achieved by having a large aperture cavity, which also provides more room for the beam to pass through.

Careful consideration was taken to select the operating RF frequency. As mentioned above, the frequency determines the length of the bucket, and this must be sufficiently long to capture the injected bunches without losing particles, even in the presence of phase and energy errors. Operating at lower frequency would result in longer buckets, but this would require larger cavities and would thus present more difficulty in applying superconducting technology. The design team settled on 400 MHz, with the corresponding bucket length of 2.5 ns (twice the frequency of the injector RF system).

Having fixed the frequency, the dimensions – in particular the diameter – of the cavity become fairly well defined. We want each cavity to control only one beam, the other should be far enough away horizontally to avoid the cavity. This means that the spacing of the two beams in Point 4 had to be increased to 420 mm from the standard 194 mm in the rest of the machine. Special magnets to separate the beams and then bring them back together again have been installed on either side of Point 4.

As the energy of the beam is increased during acceleration, the frequency must be slightly adjusted. Fortunately, the particles injected at 450 GeV are already travelling close to the velocity of light; thus, during the large change in energy that occurs during acceleration, the speed of the beam particles changes only slightly. To follow the acceleration, the RF frequency need only be modified by about 1 kHz. However, much broader tuning is an important tactic in the struggle to control other parameters, such as beam-loading effects and the minimization of cavity power. Because of this, the LHC maintains the far bigger tuning range of approximately 100 kHz.

Another design parameter is the maximum RF voltage. As we have seen, only 485 kV is required for acceleration, but much higher RF voltages are needed to achieve the minimum bunch length during storage mode to maximize beam luminosity, as well as to cover the needs for beam capture at injection. For these reasons, a much larger maximum value of 16 MV per beam is required.

Based on this value, the number of RF cavities can be determined. To minimize the cost of the RF system, this number is preferably small, but the lower limit is set by the operation of the high-power coupler, the delicate device that provides the means of transferring the RF power from the transmission line into the cold, vacuum environment of the cavity (described below). As we have seen, the losses in the SC cavity itself are negligible, but it is not possible to escape the high powers needed to control beam loading and the damping of beam errors as it enters the machine (requiring 16 MV per beam). After years of intricate development, a coupler capable of handling a power of approximately 300 kW, which means 1.2 MW in full reflection, was made. With the beam current of about 1.1 A flowing in the LHC, this power can maintain control only up to the level of 2 MV in each cavity. Hence we need eight cavities per beam.

So, in summary, the use of superconducting RF cavities frees us from the energy loss to the copper cavity itself, thus allowing more flexibility in the optimization of other design considerations. Now we turn our attention to the various parts of the RF system, looking in more detail at their design and function.

The super-conducting RF modules: the technology that interacts with the beam

Providing the accelerating voltage - the SC cavities

As mentioned in the previous section, the individual cavities are quasi-spherical with large beam apertures. Spherical cavities are employed to minimize a phenomenon called *multipactoring*. This occurs when an electron is dragged from the cavity wall by the high electric field and accelerated to a nearby surface where it may have sufficient energy to knock out more than one electron. If the time taken to do this is half an RF period, then the new electrons are in turn accelerated; this cumulative process can continue until an arc is formed, which would lead to catastrophic failure of the cavity.

The technology to produce the cavities is based on that developed for the LEP machine, a lepton accelerator which had much lower beam currents but needed exceptionally high accelerating voltages. Bare copper cavities (Fig. 92) are produced by spinning half-shells from 3-mm copper sheets, sufficient to withstand the buckling forces when the cavity is under vacuum but not too stiff to prevent tuning. The half cavities are then electron-beam welded together and coated with a niobium film, of 1-2 μm thickness, by magnetron sputtering. Niobium films, easily cooled by contact with the copper, are less susceptible to quenching (the loss of the superconducting state due to warming) than solid niobium, which is a poor heat conductor. Films are also insensitive to the earth's magnetic field, while thick superconductors trap magnetic fields when cooled; thus the use of films eliminates the need for magnetic shielding. Niobium attains its superconducting properties at 4.5 K, easily supplied by the cryogenic plants that cool the magnets to 1.7 K.

Cavity surface quality is of primary importance, any speck of dust can be an electron emitter and lead to multipactoring and a quench. Careful cleaning under stringently controlled conditions is essential. The cavities, welded inside a helium tank consisting of 2-mm-thick stainless steel, are rinsed with jets of ultra-pure water and are then transferred to Class-100 clean rooms (less than 100 particles with diameter > 0.5 microns per cubic foot!) for assembly in series of four. This is achieved with vacuum bellows which allow a small movement, without danger of vacuum leak, to take place as the ensemble shrinks during cooling. The helium tanks are designed to limit the amount of liquid helium to that strictly needed to take away the heat dissipated in the cavity.

91. Final touches to a superconducting RF module after assembly in the cleanroom (in the background).

92. A view of the bare RF copper cavities.

Tuning the cavity

Tuning is achieved by compressing the cavity along its length. With its helium tank, the cavity is very stiff – in fact, a 20-kN force is needed to change the length by 0.08 mm, the distance needed to cover the full tuning range. This means a very rigid cradle is needed around the cavity within which the cavity can be compressed (Fig. 93). The cradle also allows movement during cool-down while, at the same time, providing low-heat conductance to minimize heat loss. A specially designed lever system providing frictionless movement and eliminating backlash is moved by stepping motors.

Getting the power into the cavity - the High Power Coupler

This coupler is certainly the most delicate single element of the cavity and was one of the most challenging to make[5,6]. It provides the interface between atmospheric pressure and cavity vacuum, and a second interface between room and cryogenic temperatures. It must pass the 300 kW of power at 400 MHz (which under certain matching conditions is equivalent to 1.2 MW) and it must be moveable. The conditions at injection (where we want quick beam control and need the full available power) and in collision mode (where we want high voltages just to hold the beam lightly) require very different coupling values. These are selected by adjusting the distance that the coupler tip protrudes into the cavity. Views of the coupler are provided in Figure 94.

The heart of the coupler is a cylindrical ceramic window, brazed onto two copper rings, which provides the path for the power into the cavity. Copper is used to minimize the heating due to the high RF powers but brazing this combination of copper and ceramic is technically very challenging. The design of the copper rings themselves is critical to lowering the stresses during brazing and preventing cracking. Over 250 separate components are used in this coupler and it typically takes nine months to complete the manufacture – all items have to be made to very high tolerances. Couplers also have to be designed to prevent multipactoring; a complex conditioning procedure is employed to prevent destruction of the coupler by arcing, during which the power is raised using the vacuum pressure.

93. The cavity is held at its ends by the two square plates. A system of lever arms allows the two "blades" to be bent, compressing or stretching the cavity.

94. A schematic view of the high power coupler and the coupler waiting on the bench for installation in a cavity.

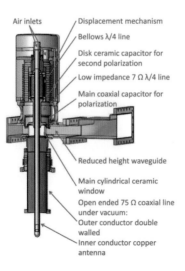

Air inlets
Displacement mechanism
Bellows λ/4 line
Disk ceramic capacitor for second polarization
Low impedance 7 Ω λ/4 line
Main coaxial capacitor for polarization
Reduced height waveguide
Main cylindrical ceramic window
Open ended 75 Ω coaxial line under vacuum:
Outer conductor double walled
Inner conductor copper antenna

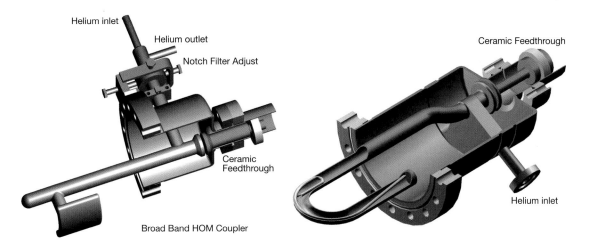

Helium inlet

Helium outlet

Notch Filter Adjust

Ceramic Feedthrough

Ceramic
Feedthrough

Helium inlet

Broad Band HOM Coupler

95. The two HOM couplers, broad band (left) and narrow band (right).

Making the cavity transparent - the HOM couplers
RF cavities resonate not only at the desired frequency, but also at higher frequencies, the higher-order modes (HOMs)[7]. High voltages can be developed at these frequencies as the beam passes through the cavities, leading to beam instability and heating of the cavities. The impedance of these HOMs has to be reduced; this is done by inserting helium cooled coupling devices matching the field configurations of these modes. Of course, the HOM couplers must not couple to the main accelerating mode. Four of these devices per cavity are necessary to cover all modes, and two are shown in Figure 95.

141

Keeping the cavity cold - the cryo-module
The cryo-module is a stainless steel cylinder with aluminum panels in its sides for easy access (Fig. 96). An assembly of four cavities with their helium tanks is rolled into the cryostat on rails and all connections to the outside world are then put in place. These connections are designed to minimize thermal losses. The power couplers are mounted last of all in a clean room. The whole cryostat is evacuated, again to reduce thermal losses. These static losses amount to

96. The cryo-module without its aluminium panels.

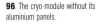

97. Safety for the cavities. The safety valve, first line of defense, and the rupture disk protect the cavity against overpressure.

98. The klystron inside its cage with the circulator and high power load in the background.

150 W per cryo-module, measured at 4.5 K. They increase by the dynamic losses of 100 W with nominal field. The total liquid helium needed to fill one module is about 320 liters.

As stated above, it is critical to keep RF noise to extremely low levels, and this is especially true for the cryo-modules. Important noise contributions come from cavity vibrations, microphonics (induced by the vacuum pumps) water cooling, helium flow and tuner movement. The design has to take account of this and to reduce wherever possible the sources. For these cryo-modules the resultant phase noise at the critical synchrotron frequency is entirely adequate for good beam-lifetime.

Safety

The pressure in the helium tank can rise dramatically in the event of a quench or in the case of a sudden loss of cavity vacuum. The helium distribution line, referred to as the QRL, is also connected to the magnets; large transient-pressure rises are possible if one or more of the main dipoles quench. To protect the RF equipment, while allowing them to quickly return to operation, non-return valves and pressure release valves are used (Fig. 97). As a final safety measure, rupture disks are placed on the He tanks. If these blow they must be replaced.

The RF power source

As we have seen, each cavity requires up to 300 kW of power at 400 MHz. To provide optimum control of the beam, we use a separate power source for each cavity, with the power supplied by 16 *klystrons* [8]. A klystron is a specialized linear-beam vacuum tube that functions as a high-gain power amplifier at radio frequencies. These devices are quite efficient with a gain of about 37 dB. Although klystrons are readily available, the LHC – in order to allow the use of fast feedback loops – had particular requirements in addition to placing the center frequency at 400 MHz; specifically, the delay through the klystron has to be less than 120 ns, and a reproducible response at higher frequencies was required. Each klystron needs approximately 54 kV with a maximum cathode current of 9 A to provide the 300 kW. Four klystrons are connected to one power converter, and all high-voltage equipment is placed in fireproof concrete bunkers. The RF conversion efficiency is about 62%, so with all other ancillary equipment included, about 8 MW of wall-plug power is required. All "excess" power has to be removed by a complex water-cooling system. A klystron can be seen in Figure 98.

99. In the underground cavern, the waveguides rise from the ground floor, where the klystrons will be placed, onto the platform and then go towards the rear of the cavern passing through a double wall into the tunnel. Prior to the RF installation this point was used to bring magnets into the tunnel – a short magnet section can just be seen in the background as it is lowered onto the bridge.

Getting the power to the cavity

The klystrons are attached to the cavities via ferrite circulators which protect the klystrons from the reflected power emanating from the cavity, each circulator having a high power RF load capable of taking full power. Waveguides connect the circulator to the cavity about 22 meters away on the other side of the tunnel wall. As the klystrons are on the floor of the cavern, the waveguides have to first rise to a platform on the level of the tunnel (Fig. 99). They then pass through rectangular holes in the double tunnel wall, there to protect the electronics in the cavern from radiation when the machine is operating and the personnel in the cavern when equipment is being tested, and arrive just above the cavities. Although the path is convoluted, care is taken to minimize the number of bends as these can be sources of breakdown in the waveguide.

Controlling the cavity and beam, system behavior

Use of multi-nested loops

The beam is controlled through the careful use of feedback loops [9]. A feedback loop operates by first measuring a parameter somewhere in the equipment or on the beam, then the measured quantity is compared to a reference value, and the system reacts by "feeding back" a correction signal to minimize the difference. The higher the gain of the loop, the better the correction becomes, but at the increased risk that the loop is more likely to become unstable. A simplified feedback loop is shown in Figure 100.

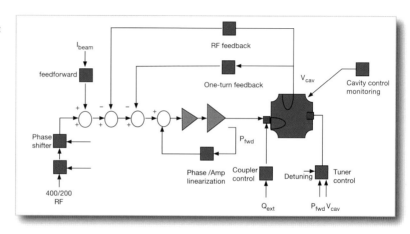

100. An example of the nested loops of the control systems, showing the one that controls the cavity parameters. Input signals are compared with references, and feedback signals are used to make and required correction.

The first feedback loop we consider is the one that controls the RF voltage in the cavity. Indeed, the RF voltage is monitored at a number of strategic points, including at the input to the RF coupler and in the cavity itself via a set-up of small probes. The feedback loop has to be particularly fast because it must be able to react to voltages induced in the cavity by the beam current. When working ideally, the loop provides an environment through which the passing beam does not "see" any induced voltage and for which the cavity will be "invisible". In addition, it helps to tightly define the amplitude and phase of the desired RF voltage. This loop also plays an important role in reducing the noise coming from the different RF components. The speed of the loop is limited by its total delay – of the order of 750 ns – which is minimized by the careful selection of components and by housing all equipment as close to the cavities as possible.

Some other feedback loops are critical to the LHC performance. One of these, with smaller gain, partially overcomes this bandwidth limitation by making corrections exactly one turn later than the measurement. This is possible as the beam reacts at its synchrotron frequency (the frequency at which the particles move back and forth within the bunch), which is much smaller than the revolution frequency.

A slow loop is used to control the cavity frequency via the tuning system by comparing the RF phases at the power-coupler input and in the cavities. The klystron loop compares the output of the klystron with its input and applies a correction via a modulator placed at the input, thus compensating against slow drifts in phase which could drive other loops towards instability. It also provides a significant reduction in noise. Since klystrons generate a lot of phase noise, the klystron loop is essential for good beam lifetime.

The loops described above do not look at the beam itself; a variety of additional loops take signals from pick-ups in the vacuum chamber and react to keep the beam stable. Two most important are the beam-phase loop and the frequency loop. The first keeps the beam stable by damping movement relative to the RF and, in the process, by reducing noise; the second keeps the beam on-orbit by ensuring that the RF frequency is correct for the magnetic field in the dipole. Close behind in importance is a longitudinal damping loop – this loop measures any large beam oscillations at injection and helps to damp them. These are supplemented by a loop to ensure synchronization with the injector.

More exotic loops

All those just mentioned above are referred to as nested feedback loops; in addition to these, we have a feed-forward loop which measures the beam current, calculates the effect on the cavity and preemptively corrects the reference signal – this improves the stability of all the nested feedback loops.

To be sure that all this is working correctly, a vast number of signals are acquired and monitored. In the event of a failure these signals are "frozen" and then given a precise time stamp to allow a "post-mortem" of the event, during which the engineers will attempt to find the cause of what went wrong.

Many of these signals come from a critical hardware element installed in each beam-line, called the longitudinal pick-up. This device detects the position of the beam longitudinally, and thus directly determines the shape and length of the bunch (Fig. 101). This measurement requires the large bandwidth of about 2 GHz, and it must be sensitive enough to supply large signals, even at low intensity, to provide noise-free references for all the loops. It works by measuring the beam current induced in the vacuum chamber walls with strip-lines placed across a cut in the chamber. These elements are placed inside a ferrite-filled vacuum-tight cylinder. Leads are fed through from the strip-lines to the exterior of the outer cylinder to allow the signals to be summed in special wide-band networks. The ferrite filling prevents the outer cylinder from shorting the gap, except at very low frequencies. The detailed shape of these ferrite components is carefully designed to provide maximum frequency response.

Transverse control of instabilities

Up to now we have been concerned mainly with the longitudinal plane, but we also have to worry about keeping the beam stable in the transverse plane, horizontally and vertically, and damping any unwanted motion, in particular at injection, that may occur [10]. Four sets of transverse dampers are installed next to the cryo-modules, two to provide a transverse electric field that acts on the two beams in the vertical direction and two others for the horizontal direction (Fig. 102). These dampers operate over a wide bandwidth, working over a frequency range of instability from about 3 kHz to approximately 20 MHz. To damp injection oscillations quickly before the beam size can in-

102. The transverse damping systems installed in the tunnel. The amplifiers sit beneath the kickers through which the proton beam passes.

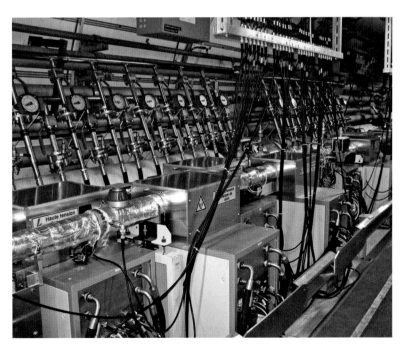

103. Looking inside the kicker as the beam sees it. In this case the two electrodes, supported on ceramic rings (pink) are at the top and bottom of the pipe to kick the beam vertically.

crease, a strong kick is applied if needed directly at the injection phase. This is the role of the transverse dampers, which can provide ± 7 kV across a gap of 52 mm and over a length of 6 m for each plane and beam.

Kickers

The 6-m length of the transverse damper is divided among 4 kickers. A kicker consists of two parallel copper electrodes, 1.5 m in length and shaped as 90° arcs, that are held 52 mm apart by brazing to three ceramic rings (Fig. 103). The ceramic elements provide both support and insulation. This rigid assembly is placed inside a machined stainless steel vacuum chamber to ensure the required electrode-positioning accuracy of ± 50 μm. Flexible copper pins connect the two electrodes to the outside world via vacuum tight connectors. The electrodes are made of copper to limit the heat rise from the beam-induced currents. Radiation and conduction cooling through the copper pins limit this temperature rise to 50 K. Although stainless steel would have been more rigid, water cooling would have been required, and this is to be avoided whenever possible in a vacuum environment.

Amplifiers and electronics

To provide the high voltages and bandwidth required to operate the transverse dampers, power amplifiers are placed under the kickers as close as possible to the electrodes. Tetrode vacuum tubes with 12 kV on the anode, one per electrode, amplify the signals to the required amplitude (Fig. 104). To get the bandwidth it is essential to reduce the stray capacitance and inductance where possible, and this is achieved by careful mechanical design of the leads, which are kept as short and separated as possible.

The instabilities and position errors are detected in beam position monitors placed at strategic points along the vacuum chamber. The signals are then

104. The high power tetrodes glowing red-hot as they are tested before installation.

treated digitally, with suitably adapted transfer functions, in much the same way as for the longitudinal loops, and then applied to the kicker amplifiers.

Fitting it all in the tunnel and the cave

As mentioned above, all the RF equipment is concentrated at Point 4, to take advantage of a large cavern, 70-m long, 16.6-m wide and 18.5-m high, that previously housed a large physics experiment, called ALEPH, in the time of the LEP machine (Fig. 105). The tunnel extends into and across the cavern on a concrete bridge built out of old LEP magnets. The bridge becomes very crowded where the RF equipment extends into the machine tunnel; at this position, all the waveguides, cryogenic piping, water cooling and related machinery have to co-exist. To protect equipment elsewhere in the cavern from radiation, the bridge was modified to include walls and a roof of 120 cm thick concrete. Waveguides and cables pass through this wall and a second shielding wall through carefully positioned holes. The roof can be removed with a crane to take out and replace a cryo-module in the event of a failure.

All the RF equipment which has to be close to the tunnel equipment, but not actually in the tunnel, is placed in the cavern. Steel platforms support the waveguide systems above the klystrons with their circulators and loads. The four high-voltage concrete bunkers are placed further into the cavern. Faraday cages (metal rooms providing shielding against external electromagnetic interference) house the sensitive electronics which work down to μW with 4 MW of power nearby. Control racks and uninterruptable power supplies are at the far end of the cavern near the lift. A second wall, 80-cm thick, has been built between the tunnel wall and the equipment to provide protection from gamma radiation during cavity conditioning, so that people can remain close to the equipment during commissioning. It also gives equipment protection from radiation produced by beam-gas collisions when the machine is running. This can cause single event upsets in the high-density electronics that do the signal processing.

 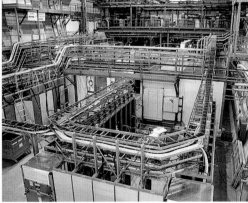

105. The UX45 cavern in 2004 (left). The bridge across the cavern to carry the beam-lines and the RF system from the tunnel on the left to the continuation on the right has been built – out of old LEP magnets! On the right, the UX45 cavern in June 2006 after installation of the infrastructure and cabling. The bridge, now with its walls and roof, is hidden behind the klystron platforms (green at the back) and a second shielding wall.

106. The RF modules after final installation in the tunnel.

The power supplies and high-voltage transformers are not located in the tunnel but rather housed, with the beam-control electronics, in a surface building. Hundreds of kilometers of coaxial, multi-wire and fibre-optic cables connect the underground areas to the surface.

The RF equipment is located far from the main control room, with a large part of it underground and inaccessible when the beam is circulating, and the complexity of the control procedures precludes direct manual intervention. Remote control is therefore essential. Switching equipment on and off, and ensuring that all parameters – cooling water flows, temperatures, powers, He-liquid pressures and levels, etc. – remain within their defined limits requires considerable computing power. Approximately 3400 signals, 1000 interlocks and 1100 alarms from the RF system alone have to be treated. At the equipment level, programmable logic controllers (PLCs) are used. These PLCs talk together and communicate to the supervisor level via an Ethernet link. The data in both these levels is treated and reduced by specific software packages before being sent to the highest level – the workstation – where high-level application programs provide the interface to the user.

The project, people and skills

This has been a 20-year project. Preliminary thoughts on the radio-frequency system for the LHC began in the late 1980s, and detailed system design got underway in the first half of the 1990s. Apart from a few research-related items, which were already needed at the earliest stages, real hardware building – and thus spending significant amounts of money – only began around 1997. At the time of writing, the last few francs of the total cost of 44 MCHF are being spent.

During this 20-year project, the work force has been continually changing. We are not the same group of physicists, engineers and technicians that started the project. Many staff from all over Europe joined an expanding CERN in the 1960s and 1970s, and many of these reached retirement age during the LHC project. There is also a natural turnover of staff but this has been fairly restricted in comparison. A constant supply of Fellows and Associates from many different countries, not necessarily those of the CERN member states, come for short periods of time, contribute to the project and then leave. There are also strong contributions from other accelerator laboratories, for example the transverse damping system described above was largely built by the JINR laboratory in DUBNA, Russia. Working with industry to build this high-tech equipment also involves close personal contact and exchange of ideas. This changing workforce, and especially contacts with different laboratories and industries, is very enriching, and new ideas are constantly available. But, at the same time, and particularly due to retirement, there is a loss of knowledge. Training and documentation ease this problem but there is usually something that was "obvious" that is not found in the books and is lost. When you finally assemble a superconducting RF cavity, only to find its frequency changes every time you cool it down, you look around for the expert! And if he has left, you have to struggle to understand the events.

During the project, RF design has become more and more exact – less of a black art – with new numerical analysis software, profiting from the ever-increasing computer power. This allows, for example, electro-magnetic design to be linked to thermal analysis and material structural behavior. Far more sophistication and certainty in the design is now possible, but this does not eliminate the need for flair and know-how, two qualities that depend on education and experience. Education in RF techniques, and especially power techniques, has diminished significantly at the university level over the past decade, and it is only recently that these gaps in the curriculum have been recognized and somewhat corrected by the offering of new courses. Finding RF engineers is difficult. Indeed, some areas in RF technology remain more of an art and rely almost exclusively on experience. For example, only certain firms can braze together unusual combinations of materials. Looking closely, one usually finds that the difference is determined by the presence of one particular technical expert who knows the tricks. Electronics design has also changed radically with the advent of digital technology. The power of digital techniques is irresistible, but to profit from this rapidly advancing technology, continual training is required. Graduates of universities and technical schools have an excellent background in these techniques, but paradoxically they lack knowledge of analogue techniques. Soon, this will be an area where action will be required.

References and further reading

[1] *LHC design report*, Volume 1, The LHC Main Ring, CERN-2004-003.

[2] D. Boussard and T. Linnecar, *The LHC Superconducting RF System*, Cryogenic Engineering and International Cryogenic Materials Conference, July 1999 Montreal, Canada and CERN LHC Project Note 316.

[3] E. Ciapala, L. Arnaudon, P. Baudrenghien, O. Brunner, A. Butterworth, T. Linnecar, P. Maesen, J. Molendijk, E. Montesinos, D. Valuch, and F. Weierud *Commissioning of the 400 MHz LHC RF System* . CERN, Geneva, Switzerland. http://cern.ch/AccelConf/e08/papers/mopp124.pdf

[4] E. Shaposhnikova, *Longitudinal beam parameters during acceleration in the LHC*, CERN LHC Project Note 242.

[5] H.-P. Kindermann, M. Stirbet, *The Variable Power Coupler for the LHC Superconducting Cavity*, Proc.9th Workshop on RF Superconductivity, Santa Fe, New Mexico, USA, 1999.

[6] E. Montesinos, *Construction and Processing of the Variable RF Power Couplers for the LHC Superconducting Cavities*, CERN LHC Project Report 1054. http://cdsweb.cern.ch/record/1080548/files/lhc-project-report-1054.pdf

[7] E. Haebel, et al., *The Higher-order Mode Dampers of the 400 MHz Superconducting LHC Cavities*, Proc. 8th Workshop on RF Superconductivity, Abano Terme, Italy, 1997.

[8] O. Brunner, H. Frischholz, D. Valuch, *RF Power Generation in the LHC*, Particle Accelerator Conference 2003 and CERN Project Note 648.

[9] P. Baudrenghien, G. Hagmann, J. C. Molendijk, R. Olsen, T. Rohlev, V. Rossi, D. Stellfeld, D. Valuch, and U. Wehrle, *The LHC Low Level RF* http://cern.ch/AccelConf/e06/PAPERS/TUPCH195.PDF

[10] P. Baudrenghien, W. Höfle, F. Killing, I. Kojevnikov, G. Kotzian, R. Louwerse, E. Montesinos, V. Rossi, M. Schokker, E. Thepenier, D. Valuch, E.V. Gorbachev, N.I. Lebedev, A.A. Makarov, S.V. Rabtsun, and V.M. Zhabitsky, LHC, *Transverse Feedback System and its Hardware Commissioning*, JINR, Dubna. http://cern.ch/AccelConf/e08/papers/thpc121.pdf

5.1

THE EXPERIMENTS

**Particle Detection
at the LHC: an Introduction**

Tejinder S. Virdee

The installation of a hadron collider in the LEP tunnel was first foreseen by CERN in the early 1980s. A Long-Range Planning Committee, set up in 1985, under the chairmanship of Carlo Rubbia, recommended that a high-energy large hadron collider, with a sufficiently large proton-collision interaction rate, was the right choice for CERN's future. To be competitive with the proposed Superconducting Super Collider in the U.S, the Committee proposed running at a center-of-mass energy of 7 TeV with a design *luminosity* of the CERN accelerator improved by an order of magnitude to 10^{34} cm^{-2}s^{-1} (leading to about one billion pairs of protons interacting per second at the heart of the ATLAS and CMS detectors). The design posed a great challenge, requiring, within a ten-year period, the development of advanced high-field superconducting dipole magnets, as well as the development of detectors capable of handling such high interaction rates. Indeed, CERN established an extensive detector research-and-development program that triggered a wide range of studies, from properties of new materials to prototype readout systems.

The experimental program for the LHC began in earnest in March 1992 with a meeting in Evian-les-Bains, on the French shores of Lake Geneva, during which the CMS collaboration presented an Expression of Interest. After a process of peer-review, ATLAS and CMS were selected in 1993 by the Large Hadron Collider Committee to proceed towards technical proposals. These proposals were requested for November 1993 and approved in 1994. ALICE, a dedicated heavy-ion experiment to search for and study the *quark-gluon plasma*, was approved in 1997; and LHCb, a dedicated experiment to study *B physics* in general – and *CP violation* in particular – was approved in 1998. The construction of these experiments began during the late 1990s.

LHC and its experiments are arguably the most complex scientific instruments ever built. The detectors are an order of magnitude more complex than previous experiments. ATLAS and CMS will have to operate in a very harsh environment created by the hundreds of billions of particles produced every second, and they will have to accurately register the passage and energies of all these particles, thus demanding huge data collection, with the transfer and processing rates on a scale greater than ever previously attempted.

The construction of the experiments has presented formidable challenges that are, at the same time, technological, engineering, organizational and financial [1-4]. Many of the technologies deployed simply did not exist fifteen years ago, and have been the result of intensive research and development work carried out in collaboration with industry. The construction of the experiments has also required the pooling of the resources and talent of a very large number of scientists. For example, each of the ATLAS and CMS collaborations comprise over 2500 scientists and engineers from about 180 institutions in over 35 countries; ALICE and LHCb are each about a factor three smaller. The experiments are presently (as of Spring 2009) ready for the next round of collisions, now anticipated for the last quarter of 2009.

A chapter on detector basics

The prime motivation for the Large Hadron Collider (LHC) is to elucidate the nature of *electroweak symmetry breaking* for which the Higgs mechanism is presumed to be responsible (see Chap. 2). There are hopes for discoveries that might reveal physics beyond the predictions of the present theory; for instance we might observe signs of supersymmetry or extra dimensions, the latter potentially requiring the modification of gravity at the Tera-electron-volt (TeV) scale. Overall, the TeV energy scale appears to have a special significance; each of the LHC experiments is designed to study physics at these energies.

This Chapter 5 is organized into six parts. The four following this brief introduction discuss the detectors in some detail; indeed, the reader will find a number of common concepts among the basic measuring capabilities of the LHC detectors. This is not an accident – the selection of technology has been driven by the physics goals, and this introductory chapter will show how the requirements defined by the physics can be related to a corresponding set of experimental criteria. After presenting each of the four experiments in more detail in the subsequent chapters, the final one of this section will present some of the aspects of data handling, which, due to the magnitude of this challenge has been referred to as the "fifth experiment."

Detectors for hadron colliders – the last 30 years

One of the first high-energy colliders to come into operation was the CERN Intersecting Storage Rings (ISR) in the early 1970s. The experiments at the ISR aimed at studying only small regions of solid angle with respect to the collision point, rather than completely surrounding the collision with a fully instrumented device, as is case with many modern detectors. The luminosity at the ISR eventually reached 10^{32} cm^{-2} s^{-1} (corresponding to about 5 million inelastic interactions per second), and much was learnt about experimentation at hadron colliders. Following this, in each successive generation of collider experiments, innovative instrumentation has had to be introduced. In the UA1 detector of the ISR, the *hermetic geometry* around the interaction point was introduced for the first time; this configuration is now the standard. It also pioneered on-line triggering using the full event information in order to decide which events to keep for future analysis. Another feature was the gaseous central tracking detector, immersed in a magnetic field, which gave electronic "bubble-chamber-like" images. The technological concept of *high-granularity calorimetry* was introduced by UA2.

Later, starting in 1989, the LEP collider at CERN was built with combinations of superconducting solenoids; *micro-vertex detectors* to detect b-quarks; bubble-chamber-like tracking chambers such as *time-projection chambers* (TPC); high-granularity electromagnetic (EM) calorimeters; and ring-imaging *Cerenkov counters* (RICH) for particle identification. In the following pages, these ground-breaking techniques will be presented in terms of their role in LHC detection, for which the harsh conditions in a high-luminosity environment required these different detector technologies to be further developed and modified to meet the additional challenge of increased radiation stresses.

The experimental challenge at the LHC

When the LHC performs according to its design, the general-purpose detectors (ATLAS and CMS) will observe 6.5×10^8 inelastic proton-proton collisions per second. This leads to a number of formidable experimental challenges [1].

The event selection process (called the *trigger*) must reduce the billion interactions that occur each second to no more than few hundreds of events for storage and subsequent analysis. The short time between bunch crossings, 25 nanoseconds (1 ns = 10^{-9} s), has major implications for the design of the readout and trigger systems. It is not feasible to make a trigger decision during these 25 ns, yet new events occur in every crossing and a trigger decision must therefore be made for every crossing; the solution is to split the decision in three levels and to store events that are good candidates for interesting physics in a temporary memory during the decision process. The first of these is the Level-1 trigger decision, which takes about 3 μs. During this time the data must be stored in *pipelines*.

At the design operation, on top of the selected or otherwise interesting event, an average of 20 other proton-proton events will be superimposed. These 20 or so are referred to as *minimum-bias* events, because no selection is made, i.e., no bias is introduced for these events. However, this implies that around 1000 charged particles will emerge from the interaction region every 25 ns. The products of an interaction under study may be confused with those from other interactions in the same bunch crossing. This problem, known as *pileup*, clearly becomes more severe when the response time of a detector element and its electronic signal is longer than 25 ns. The effect of pileup can be reduced by using highly granular detectors with good time resolution, giving low *occupancy* (i.e., the probability that a detector will return a non-zero signal) at the expense of having large numbers of detector channels. The resulting millions of electronic channels require very good time synchronization.

To give the reader an idea of the magnitude of the task, the online trigger system has to analyze information that is continuously generated at a rate of 40,000 GB/s and reduce it to an order of one GB/s for storage. Even after this tremendous reduction, tens of petabytes (1 PB = 1,000,000 GB) of data will be generated annually by each experiment and distributed for offline analysis to scientists located around the globe. This data management problem motivated the development of the so-called "Computing Grid" (Chap. 5.6).

The large flux of particles emanating from the interaction region creates a high-radiation environment requiring radiation-hard detectors and front-end electronics. Access for maintenance will be very difficult, time consuming, and highly restricted. Hence, a high degree of long-term operational reliability has to be attained, comparable to that which is usually associated with instruments flying on space missions.

The two "smaller" detectors have to face different but equally difficult problems. The main challenges for LHCb detector are the efficient online selection of events containing b-flavored hadrons; particle identification; and the precise determination of the positions of particle interactions (*vertexing*). For ALICE, which will focus on heavy-ion collisions, the main challenge will be to handle extremely complicated events containing tens of thousands of tracks.

Thus, the LHC general-purpose detectors (ATLAS and CMS) and the specialty detectors (LHCb and ALICE) cannot simply be larger versions of the previous generation of HEP detectors. A major research-and-development effort was required to develop detectors and electronics that could survive and operate reliably in the harsh LHC environment.

Analogy of an HEP detector with a cylindrical onion
High energy physics detectors, such as ATLAS and CMS, are comprised of several layers (see Panel 1). Each layer is designed to perform a specific task, and together they allow identification and precise measurement of the energies of all of the particles produced in LHC proton-proton collisions. The layers of the detector are arranged like a cylindrical onion around the collision point. A transverse section of the CMS detector illustrates this in Figure 1 of Panel 1.

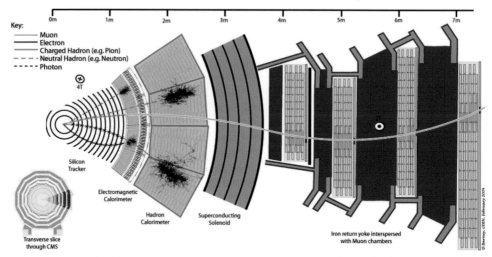

1. A schematic of a slice of the transverse cut through CMS illustrating how the identity of particles produced in proton-proton collisions is ascertained.

Panel 1
Pattern Recognition: Identity of Particles
New particles discovered at the LHC will be typically unstable and rapidly transform into a cascade of lighter, more stable and familiar particles. Particle traveling through the LHC detectors leave behind characteristic patterns, or "signatures", in the different layers, allowing them to be identified.

We now look at some of the typical signatures that we expect using a cross section of the CMS detector in the illustration above. Muons are penetrating charged particles that leave 'hits', that are linked up to create 'tracks' in the inner tracker and the muon tracking system but little energy in the calorimeters. Electrons are charged particles that leave tracks in the inner tracker and deposit all their energy in the electromagnetic calorimeter and nothing beyond. Photons are neutral particles and hence leave no tracks in the inner tracker but otherwise behave like electrons when coming in contact with the calorimeter. Charged hadrons such as pions leave tracks in the inner tracker, some energy in the electromagnetic calorimeter and the rest in the hadron calorimeter and nothing beyond. Neutral hadrons do not leave tracks in the inner tracker but otherwise behave like charged hadrons upon contact with the hadron calorimeter. Neutrinos are highly penetrating neutral particles, even capable of traversing the earth, whose presence is inferred from the mismatch of momenta in the transverse plane (perpendicular to the beamline).

A particle emerging from the collision and travelling outwards will first encounter the tracking system immersed in a strong solenoidal field. The system consists of *silicon pixels* and *silicon microstrips*, or gaseous *straw-tube* detectors. These detectors precisely measure the paths of charged particles, allowing physicists to reconstruct their trajectories. Charged particles follow spiraling trajectories in the magnetic field, and the curvature of their trajectories reveals their momenta, from which a first measurement of the energy of the charged particles can be derived. The energies of all the particles, including the neutrals, will be measured in the next layer of the detector – the so-called *calorimeters*.

The first calorimeter layer is designed to measure the energies of electrons and photons with great precision. Since these particles interact mainly electromagnetically, it is called an *electromagnetic calorimeter*. Particles that mostly interact by the strong interaction, called hadrons, deposit most of their energy in the next layer, in the *hadron calorimeter*. The only particles to penetrate beyond the hadron calorimeter are muons and neutrinos. Muons are charged particles that are tracked in the outermost dedicated muon chambers, also immersed in a magnetic field so that their momenta can be measured from the curvature of their trajectories. Neutrinos, however, are neutral; since they hardly interact at all they will escape direct detection. Their presence can nevertheless be inferred in the following manner: since the total vector sum of the momenta of the particles is conserved during the collision, the total momentum along the x and y axes (taking the y axis to be perpendicular to the beam axis) has to be zero. By adding up the momenta of all the detected particles, physicists are able to tell where the neutrinos went and how much momentum they took away.

The experimental magnets

The single most important aspect of the overall detector design is the choice of the magnetic field configuration for the measurement of muons, as this decision strongly influences the rest of the detector design. The two basic configurations are *solenoidal* and *toroidal*, characterized by magnetic fields with direction parallel and perpendicular to the beam axis, respectively. Large bending power is needed to precisely measure high momentum muons or other charged tracks. CMS chose a superconducting high-field solenoid, with a large length-bore ratio; whereas ATLAS chose a superconducting air-core toroid. The CMS solenoid, in addition, provides the magnetic field for the inner tracking system whilst ATLAS has an additional solenoid magnet to carry out the same function. The very different basic structures of these two magnets can be seen in Figures 2 and 3.

2. Photograph of the CMS solenoid magnet during construction. Additional views of the solenoid are found in Fig. 16.

3. Image of three of the ATLAS toroids during the construction phase of the experiment. The completed toroidal magnet is seen in Fig. 28.

Inner Tracking System

The most powerful way to 'see' the event topology is by using the *inner tracker*. The inner tracking systems are designed to provide a precise and efficient measurement of the trajectories of charged particles emerging from the LHC collisions, as well as a precise reconstruction of *secondary vertices* (i.e., the position where a particle produced in a collision itself further decays into a new set of particles).

The ATLAS and CMS tracking detectors at the LHC have to deal with very high particle fluxes (about 2×10^{10} particles per second emerging from the interaction point) and a very short time between bunch crossings. In addition, we need to measure the momentum of energetic charged particles (100 GeV/c) with a resolution of approximately 1%; this represents an order-of-magnitude improvement over the LEP experiment. The very precise silicon-pixel and microstrip detectors, along with short-drift-time gaseous detectors (straw tubes) are the technology of choice.

The particle flux decreases as the observer moves away from the particle source, varying from 10^8 cm^{-2} s^{-1} at a radius of $r = 4$ cm; to 5×10^6 cm^{-2} s^{-1} at $r = 25$ cm; and 5×10^5 at 100 cm. Thus, three basic regions can be defined (Fig. 4):

- At radii between 4 and 20 centimeters, there are several layers of 'hybrid' pixel detectors. The typical area of a pixel is 10^4 μm^2 giving an average particle *occupancy* (average number of detectors having registered a signal) of about 10^{-4} per pixel per LHC bunch crossing.
- In the intermediate regions (at radii between 20 and 60 cm) the particle flux is low enough to enable use of silicon microstrip detectors with a typical cell size of 10 cm $\times$ 75 μm leading to an occupancy of 1% per LHC crossing.
- In the outermost regions of the inner tracker, the particle flux drops sufficiently to allow use of larger pitch silicon microstrip (CMS) or gaseous straw tube detectors (ATLAS). The typical cell size in CMS is 25 cm $\times$ 180 μm, which results in an occupancy of a few percent per crossing.

4. Two views of the inner tracking systems used at the LHC: (a) a schematic diagram of the inner tracking system of the ATLAS experiment showing the pixel detectors close to the beam axis, followed by layers of the silicon microstrip detectors (SCT) and finally the straw-tube detectors of the TRT (described in Chap. 5.3); and (b) a photograph of the inner tracking system of CMS showing three layers of silicon modules. Positioned close to the interaction point of the proton-proton collisions, the silicon used here must be able to survive high doses of radiation and a 4-T magnetic field without damage.

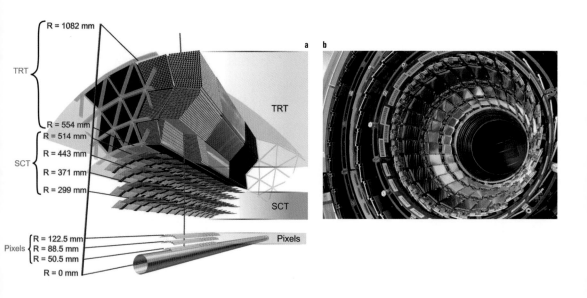

5. A technician places a lead tunstate scintillating crystal in place for the ALICE calorimeter, which uses the same crystal technology as the CMS experiment for its electromagnetic calorimeter.

Electromagnetic calorimeter

The electromagnetic calorimeters in ATLAS and CMS face one of the greatest scientific challenges of the LHC, namely the detection of the two-photon decay of an intermediate-mass Higgs boson. The signal would be observed as a peak in the reconstructed mass of the object that generated the two photons; the background is large, and the only hope of finding the Higgs signal is if the accuracy in mass resolution is very good, so the signal peak is sharp, high and narrow.

For its electromagnetic calorimeters (and also for the end-cap hadronic calorimeters), ATLAS uses liquid argon as the active medium. Calorimeters using liquid-filled ionization chambers as detection elements have several important advantages. The absence of internal amplification of charge results in a stable calibration over long periods of time. The considerable flexibility in the size and the shape of the charge-collecting electrodes allows high granularity both longitudinally and laterally. ATLAS introduced a novel absorber-electrode configuration, known as the "accordion" in which the particles traverse the chambers at angles around 45°.

CMS chose lead tungstate scintillating crystals ($PbWO_4$) for its electromagnetic calorimeter because active calorimeters (such as these inorganic scintillating crystals) provide the best possible performance in terms of energy resolution. These crystals offer several advantages for operation in CMS, including compactness; a length of only 23 cm is sufficient to absorb the energy of the high-energy electrons and photons produced during the proton collisions.

The hadron calorimeter

Hadron calorimeters, in conjunction with the electromagnetic calorimeters, are primarily used to measure the energies of *jets*, which are collimated bunches of particles originating from the fragmentation of quarks and gluons. The most important characteristics for the performance of hadron calorimeters are
- jet-energy resolution and energy linearity;
- hermetic coverage for measuring missing transverse energy resolution.
  By measuring the energy of hadrons, this class of calorimeter provides an

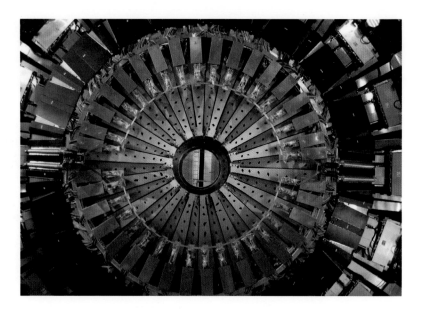

158

indirect measurement of the presence of weakly interacting particles such as neutrinos.

The ATLAS hadron calorimeter uses alternating layers of steel 'absorber' and liquid fluorescent 'scintillator' such as liquid argon. When a hadron (e.g., a proton or charged pion) hits an iron nucleus, numerous secondary particles are created which in turn strike further iron nuclei. A cascade of particles thus develops. The charged particles traversing the scintillator generate light in the scintillator, or free charges in the liquid, the amount of which is proportional to the deposited energy and hence also to the energy of the initially incident hadron.

The CMS hadron calorimeter uses a different technology. It consists of a brass/scintillator-tile sampling calorimeter with fine lateral granularity. The scintillation light is observed by wavelength shifting fibers, with the light channeled on clear fibers to hybrid photo-diodes placed axially to the 4-tesla magnetic field.

The muon system

We anticipate that high momenta muons will be produced in the decays of certain as-of-yet-undetected particles. For example a standard-model Higgs boson can decay into two Z bosons; each of these can, in turn, decay into a pair of muons, leading to four muons traversing the CMS detector. A powerful muon system is therefore needed to identify, trigger, and accurately measure the momentum of these muons. Good muon-momentum resolution and triggering capability are aided by a high magnetic field, either in air (as in ATLAS) or in a thick flux-return iron yoke (as for CMS). The latter also serves as a hadron filter that improves the identification of muons.

Two kinds of highly complementary muon detectors are used at LHC. These are the *gaseous drift chambers* (GDC) that provide accurate measurement in space (position) needed for accurate determination of momentum; and the "trigger" chambers, such as *resistive plate chambers*, that provide an accurate measurement of the time of passage of the muon. This enables the muons to be assigned to the correct event. At the LHC, the number of hits recorded in the end-cap chambers (≈ 1 kHz/cm^2) is two orders of magnitude

7. Installation of the barrel muon detector of the CMS showing how the muon system is structured with alternating layers of iron return yoke (red) and muon detection chambers.

larger than in the barrel region. Therefore, drift-chambers are replaced by detectors with a faster response, such as cathode strip chambers.

Electronics Chains for LHC Experiments

A very substantial part of the material cost of the LHC experiments was invested in electronics, some of which has to be *radiation hard*, previously a requirement typically found only in military and space applications.

The features that differentiate the electronics of the LHC experiments from previous ones are the need for high-speed signal processing; the presence of large *pileup* (mentioned above); the ability to function in high radiation environments; the far larger number of channels (larger data volume); and the application of in-house designed electronics using state of the art technologies.

The Experiments

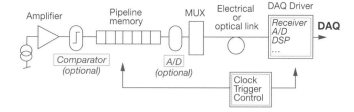

8. A generic readout system for a detector at the LHC.

A generic LHC readout system is illustrated in Figure 8. Indeed, there are a number of functions that are common to almost all systems:

- amplification;
- analogue to digital conversion;
- association to beam crossing;
- storage prior to trigger;
- dead-time-free transmission via optical links;
- data reduction via removal of empty signals (zero-suppression); and
- formatted storage prior to access by the data acquisition, calibration, control and monitoring systems.

All the electronics inside the experiment cavern have to be *radiation tolerant* whereas those in the high radiation environment (inner tracker, ECAL etc.) have to be *radiation hard*. This has meant that an unusually large fraction of the electronics has had to be custom-designed – a new challenge in high-energy physics. As the minimum feature size for electronics has become smaller, the cost per chip and the cost per channel has decreased. In the last decade considerable effort has gone into designing high-performance, low-power-consumption, low-cost-per-channel and radiation-hard electronics.

Selection and acquisition of interesting events: ATLAS and CMS as 100 megapixel 3-D digital cameras taking 40 million shots per second!
ATLAS and CMS each have about 100 million electronics channels with a timing resolution of nanoseconds; the amount of data produced will be enormous. It is interesting to think of each of these detectors as a 100-mega-pixel 3-D digital camera. These are of course no ordinary cameras – to start with, the CMS weighs around 14,000 tons – and each one takes 40 million pictures per second, representing roughly the number of times bunches of protons cross each other in the heart of the detectors. We anticipate that about 20 pairs of protons will interact during each crossing, of which even fewer will contribute to the picture of interest (particle physicists call these 'events'). Each event produces around 1 MB of data, leading to a generation of around 40 Terabytes (1 TB = 1000 GB) of data each second. However it is not possible to transfer, analyze and store this much data. Hence a small subset of this information, with coarse granularity, is sent out of the detector on optical fibers to an adjacent underground cavern for analysis by the off-detector electronics. Dedicated processors select about 100,000 of the most interesting events on the basis of large energy deposits as they transverse to the beamline. The data from these selected events, amounting to 100 GB per second, are transferred to the computing facilities on the surface via about 1000 optical fibers. The data from each event is sent to the next free CPU in a farm of about 30,000 cores. The same physics software code runs in each one of these cores to select about one in a thousand, using the full granularity and resolution information, as well as the desired physics quantities and signatures. These data, amounting up to one GB per second (100-200 events), are sent out to the CERN central computing facilities; the story of what happens to the data at this point is the subject of Chapter 5.6.

Panel 2
Two General Purpose Experiments
Down the ages the *scientific method* has relied on scientific observation and independent confirmation. Hence when searching for any new phenomena, especially in such complex environments as at the LHC, two independent experiments, based on different approaches, is invaluable. Each of the LHC experiments uses cutting-edge technologies in the detector design; it is also important to point out that these technologies are also complementary. For example, the detection and measurement of muons in one experiment relies on a magnetic field generated by superconducting toroids, whilst the other one uses a superconducting solenoid; for the inner tracking, one experiment uses a combination of silicon and gas detectors whilst the other one uses only silicon detectors; for the measurement of the energies of electrons and photons one experiment uses a technology based on liquid argon whilst the other one uses scintillating crystals, etc. The differing technologies have quite different *systematic errors* (a quantity that encapsulates any lack of understanding of the response of an instrument) so that if both ATLAS and CMS 'see' a signal then our confidence that it is truly there.

Future developments at the LHC: towards the Super-LHC
Thoughts are already turning to the long-term future of the LHC and to ways of expanding the range of experimental conditions that might lead to new insights in physics. Notably, there is discussion to work towards a further ten-fold increase in luminosity (to 10^{35} cm^{-2} s^{-1}). This would require considerable and fresh research and development in detector technologies, including systems aspects such as cooling and power distribution, especially for the inner trackers. As currently experienced with the LHC detectors, lead times for the development and deployment of complex detectors are long (over ten years). Hence, it is not too soon to think about future development of the LHC, even though the main effort today must be devoted to finishing the current LHC construction projects.

The mystery of the TeV energy range

Much research, development and prototyping was carried out during the 1990s to develop detectors that would be able to cope with the harsh conditions anticipated in the LHC proton-proton experiments. These are not just bigger versions of the recent or currently running detectors. They are substantially different, innovative and at the frontier of various technologies. Their construction has taken place on a scale previously unseen in experimental science, with the integration of industrial production techniques for certain aspects of the individual sub-detectors. The LHC detectors have been commissioned and are ready for proton-proton collisions at high energies. LHC experiments should be capable of discovering whatever Nature has in store at the TeV energy scale and are likely to provide answers to some of the most important open questions in physics.

161

Acknowledgements

The construction of the large and complex experiment such as those at the LHC would not have been possible without the effort and the dedication of thousands of scientists, engineers and technicians worldwide. By the time the first suite of results are published, many of them will have spent a substantial fraction of their professional lives on these experiments. This paper is dedicated to the efforts of all these people and the agencies that funded them.

The Experiments

References
[1] N. Ellis and T. S. Virdee, "Experimental Challenges in High Luminosity Collider Physics," *Ann. Rev. Nucl. Part. Sci.* **44** (1994) 609.
[2] T. S. Virdee, "Detectors at LHC," *Physics Reports* **403-404** (2004) 401.
[3] D. Friodevaux and P. Sphicas, "General-Purpose Detectors for the Large Hadron Collider," *Ann. Rev. Nucl. Part. Sci.* **56** (2006) 1.
[4] ALICE: K. Aamodt et al., *JINST* **3** (2008) S08002; ATLAS: G. Aad et al., *JINST* **3** (2008) S08003; CMS: S. Chatrchyan et al., *JINST* **3** (2008) S08004; LHCb: A. Augusto Alves Jr. et al., *JINST* **3** (2008) S08005.

5.2

THE EXPERIMENTS

The Compact Muon Solenoid Detector at LHC

Tejinder S. Virdee

In several of the earlier chapters of this book, we presented the strong motivation for the exploration of the Tera-electron-volt (TeV) energy scale, as well as the role that the Large Hadron Collider will play in this grand scientific adventure. The Compact Muon Solenoid (CMS) is one of the two general-purpose experiments built to explore the physics at this scale.

The construction of CMS [1, 2] has presented formidable challenges that are at the same time technological, engineering, organizational and financial. Many of the technologies employed in CMS simply did not exist fifteen years ago; their development has been the result of intensive research and development carried out in collaboration with industry.

The CMS Collaboration involves over 2,500 scientists and engineers from over 180 institutions in 38 countries. It has taken almost two decades from the first ideas to CMS being ready for beam in the last quarter of 2008. It has also required the pooling of the resources and talents of a very large number of scientists.

The design of the CMS

As mentioned in the previous chapter, the single most important aspect of the overall detector design is the choice of the magnetic field configuration for the measurement of muons, as this aspect strongly influences the rest of the detector design. The two basic configurations are *solenoidal* and *toroidal*. Large bending power is needed to precisely measure high momentum muons or other charged particle tracks. As its name implies, CMS chose a superconducting high-field solenoid.

CMS has the classical form of a high-energy particle detector; its overall layout is shown in Figure 9. At the heart of CMS sits a 13-m-long, 5.9-m-wide, 4-tesla superconducting solenoid. In order to achieve good momentum resolution within a "compact" spectrometer, without making overly stringent demands on muon-chamber resolution and alignment, we decided to operate under a high magnetic field. The field is returned through 1.5 meters of iron structure, which houses four *muons stations* to ensure robustness and full geometric coverage. Each muon station consists of several layers of *aluminium drift tubes* (DT) in the barrel region and *cathode strip chambers* (CSCs) in the endcap region, complemented by *resistive plate chambers* (RPCs). These elements are discussed in more detail below.

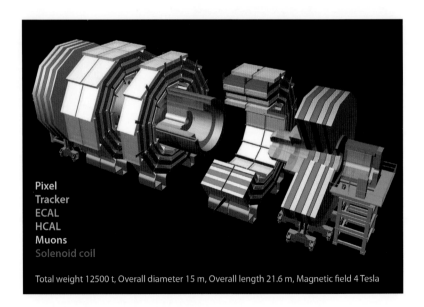

Pixel
Tracker
ECAL
HCAL
Muons
Solenoid coil

Total weight 12500 t, Overall diameter 15 m, Overall length 21.6 m, Magnetic field 4 Tesla

164

Figure 9 shows the overall structure of CMS, with the bore of the magnet coil designed to accommodate the *inner tracker* and the *calorimeters*. The cylindrical inner-tracking volume is 6 m in length and 2.6 m in diameter. In order to deal with high-track multiplicities, the CMS inner tracker employs ten layers of *silicon microstrip detectors* which provide the required granularity and precision. *Silicon pixel detectors* placed close to the interaction region improve the measurement of the impact parameter of charged-particle tracks as well as the position of secondary vertices. The particles then encounter the calorimeters, firstly the *electromagnetic calorimeter* (ECAL), which uses lead tungstate ($PbWO_4$) crystals for the measurement of the energy of photons and electrons. The ECAL in turn is surrounded by a brass/scintillator-sampling *hadron calorimeter*. The *forward calorimeters* ensure full geometric coverage for the measurement of the transverse energy in the event.

The construction of CMS

Before discussing the instrument itself in detail, we present some of the remarkable challenges overcome during the construction of this massive detector, which was built in modules and subsequently assembled inside a cavern located 100 meters below the surface of the Franco-Swiss plain in the region known as the Pays de Gex.

At the conceptual design stage, some 20 years ago, it was decided to divide the massive flux-return iron yoke of the solenoid into sections. This decision allowed construction and assembly in stages. The yoke was sectioned into five barrel-wheels and four endcap-disks on each side, visible in Figure 9 (with only three initially installed for the low luminosity detector). The strategy called for the assembly of the yoke in a large surface building specially conceived for this purpose (Fig. 10). It is of interest to note that the muon chambers and the hadron calorimeter were also installed and tested in the same surface building (Fig. 11).

One of the more remarkable pieces of construction-site equipment was the large gantry crane constructed on the outside of the surface building to lower the heavy elements of the CMS detector in 15 pieces. The largest element

10. Aerial view of the CMS construction site, showing the building used for the construction of the yoke.

11. A inside view of the surface assembly hall of CMS in 2006 during the assembly of CMS. The sectioned structure of the yoke is apparent.

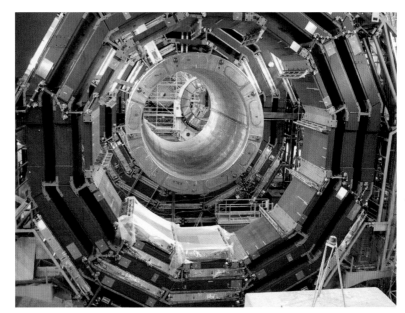

12. The lowering of an element of the CMS using the gantry.

The Experiments

13. Inner view of the CMS showing the complexity of the equipment housed inside the solenoid.

was the central wheel that housed the solenoid, a piece weighed in at 1,920 tons (Fig. 12). For the lowering operation, elements were suspended by four massive cables, each with 55 strands, and attached to a step-by-step hydraulic jacking system, with sophisticated monitoring to ensure the object did not sway or tilt. Each lowering operation took around 10 hours. The first heavy element was lowered in November 2006 and the final fifteenth one in January 2008.

Indeed, the CMS experiment is the first of its kind to be constructed above ground and then lowered, piece by piece, 100 meters below. There are many advantages to planning an experiment in this way; time was saved by working simultaneously on the detector while the experimental cavern was being excavated. There were also fewer risks working at the surface in a more spacious environment. The various elements of detector could be tested together before lowering. When safely positioned in the cavern, the complex instrumentation was connected to its service sources (cooling power, optical fibers, etc.) and to the other elements of CMS; Figure 13 illustrates the enormity of the task, showing the many cables, optical fibers, cooling pipes, etc., for detector elements that are housed inside the solenoid (namely the barrel hadron calorimeter, the barrel electromagnetic calorimeter and the inner tracker).

Once the CMS is closed in its final position for data-taking, access for maintenance will be very difficult, time consuming, and highly restricted. Hence, a high degree of long-term operational reliability – usually associated with space-bound systems – has had to be attained. Nevertheless, maintenance will undoubtedly be required, in which case the modular design of CMS will be appreciated.

The following sections describe the layers of CMS particle detection, placing an accent on the particular challenges that had to be overcome, as well as on the innovative technology employed by the CMS team.

Panel 1
Construction of the Coil Modules

The construction of the CMS solenoid was the fruit of an impressive international collaboration; so much so that the radial size of the coil was limited by European regulations of transport. The NbTi superconducting wire was made in Finland, braided into superconducting "Rutherford cable" in Brugg, Switzerland and co-extruded with pure aluminum near Neuchatel, Switzerland. This "insert" was then electron-beam welded, in France, onto two plates made of the high-strength aluminum alloy that enabled the conductor to support the massive outward pressure exerted by the magnetic field. The cable, the insert, and the final conductor were made in continuous and perfect lengths of 2.6 km, a remarkable feat of engineering. The five coil-modules were wound in Italy and individually shipped from Genoa to Marseille, transported up the Rhone river and finally by truck from Macon to CERN. The 7.2-m outer diameter of the coil for CMS was chosen specifically to be small enough that it could be transported from the manufacturing site to CERN without having to widen roads or pull down and then rebuild any bridges!

The CMS superconducting solenoid magnet

As mentioned above, the choice of magnet is the driving feature of a particle detector; the CMS team chose to employ a superconducting solenoid magnet. With its maximum magnetic field of 4 tesla, and its coil design sectioned into five "modules" each of a length of 2.5 meters, the solenoid is a marvel of modern technology. For this magnet, the *magnetic flux* is returned through a 12,000-ton yoke consisting of five wheels and two endcaps, the latter composed of three disks each; the completed yoke is shown in Figure 14.

The construction of a superconducting solenoid magnet presented a number of major challenges. Specifically, several key parameters – such as the magnetic field, the number of Ampere-turns, the forces generated, and the stored energy (Fig. 15) – necessitated major changes from earlier designs. The distinctive feature of the

167

The Experiments

14. The magnetic yoke before descent into the CMS cavern.

15. The magnetic-field energy stored in the CMS

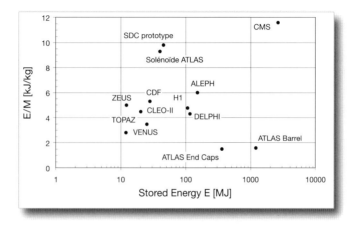

16. Some views of the solenoid coil: (a) a cross-section through the prototype coil module showing reinforced conductor and the 4-layer winding; (b) the five coil modules connected in the surface hall of CMS; (c) the rotating of the coil before insertion; and (d) the completed coil ready for the summer 2006 test.

220-ton cold mass is the 4-layer winding (Fig. 16a) made from a reinforced Nb-Ti superconductor through which some 20,000 Amperes of current flow. This novel conductor has to withstand an outward pressure of 60 atmospheres! The stored energy is very large, inducing significant mechanical deformation during energizing, well beyond the values for solenoids in previous experiments.

CMS magnet construction began in 1998, and by 2002 the fabrication of the superconducting wire was complete. The winding of the five modules of the solenoid coil began in 2000 and took five years to accomplish. The coil-modules were assembled in the vertical position and connected in the surface assembly building of CMS. After being brought to the horizontal position using a swiveling tool (Fig. 16b), the cold mass was inserted inside the outer vacuum tank from which it is suspended by titanium tie-bars. After swiveling, using the same tooling, the inner vacuum tank was inserted inside the coil (Fig. 16c), and the two flanges welded to complete the vacuum enclosure. By the end of 2005, the solenoid (Fig. 16d) was ready for testing, and in February 2006, it was cooled down to its operating temperature of approximately –267 °C. The cool-down, from room to operating temperature, takes about one month. In mid-2006, the installation of muon chambers was interrupted, and the experiment closed to test the magnet and the sections of the already installed detectors.

The world's largest superconducting solenoid magnet reached the full field in September 2006. Based on cryogenic, electrical and mechanical tests, the coil fulfills all the specifications and is, at present, fully operable.

a

b

c

d

Silicon-strip inner-tracking system

As discussed in the previous chapter, the first line of detection around the colliding protons is a tracking system for charged particles. The inner tracking system of CMS is designed to provide a precise and efficient measurement of the trajectories of the charged particles emerging from these collisions, and it precisely reconstructs the positions of the *secondary vertices*. The momentum of 100 GeV/c charged particles is measured to a precision that is improved by an order of magnitude compared to the previous generation of detectors.

The CMS tracker surrounds the interaction point, with a length of 5.8 m and a diameter of 2.5 m, bathed in the 4-T homogeneous magnetic field provided by the solenoid. At the LHC design luminosity, there will be on average 1000 particles, from more than 20 overlapping proton-proton interactions, traversing the tracker for every bunch crossing (about every 25 ns). Detector technology featuring high granularity and fast response to the passage of the charge particles is required, such that the particle trajectories can be identified reliably and attributed to the correct bunch crossing. These features require fast response and lead to a high number of channels, implying detection electronics with a high power density, which in turn require efficient cooling. These specifications have to be balanced against the aim of keeping the amount of tracker material to a minimum in order to limit a number of undesirable phenomena, such as *multiple scattering* (i.e., interaction with detector materials), *bremsstrahlung*, photon conversion, and nuclear interactions. The intense particle flux leads to high radiation levels in the tracking-system environment.

With this in mind, the main challenge in the design of the tracking system was to develop detector components and service systems able to operate

17. One "shell" of the tracker inner barrel with silicon modules in the process of being mounted.

in this harsh environment for a lifetime of over ten years. These requirements on granularity, speed and radiation hardness led to a tracker design entirely based on silicon detector technology.

The final design of the CMS tracker consists of a *pixel detector* with three "barrel" layers at radii between 4.4 cm and 10.2 cm, and a *silicon-strip tracker* with 10 barrel-detection layers extending outwards to a radius of 1.1 m. (Fig. 17). The barrel system is completed by end caps also relying on silicon pixel detectors and strip trackers on each side, extending the detection of the tracker to very small solid angles relative to the beam axis. With about 200 m^2 of active silicon area, the CMS tracker is by far the largest silicon tracker ever built.

The construction of the CMS tracker, composed of 1,440 pixel- and 15,200 microstrip-detector modules, required the development of production methods and quality control procedures that are new to the field of particle physics. Indeed, these developments are so critical to the operating concept of CMS that we now take a closer look at their development and assembly.

A Late Change of Sensor Technology - Silicon Microstrip Detectors

Silicon microstrip detectors are perhaps the ideal tracking detectors for proton-proton experiments at the LHC. They are fast (the charge can be collected within the inter-bunch crossing time of 25 ns) and can give very good spatial resolution and a fine two-track resolution (i.e., the ability to separate individual tracks inside jets of particles). At the design stage of the general-purpose proton-proton detectors (early 1990s), radiation damage of silicon was poorly understood, and the cost of sensors and electronics appeared to be prohibitively large.

Considerable research and development took place to (1) improve the understanding of the damage mechanisms; (2) work out strategies to prolong the useful lifetime of irradiated detectors; and (3) improve the high-voltage behavior of the sensors. With careful processing and use of *multi-guard-rings* it is now possible to produce sensors that can withstand very high bias voltages. Upon irradiation at levels anticipated at the LHC, a sizeable *leakage current* is induced, but more seriously, the effective doping of the bulk changes, leading to *type-inversion* (n-type material becomes p-type) requiring the use of progressively higher bias voltage to keep the detector operational. It was found that these doping changes continue even after termination of irradiation when the detectors operate at room temperature (referred to as *reverse annealing*). The CMS silicon tracker is run at −20°C to arrest the reverse annealing.

Following these developments, along with improvements in fabrication and automation, the CMS collaboration made a bold decision to change the design of its inner tracker in 1999. Originally, it had included both *microstrip gas chambers* (MSGCs) and silicon sensors; as the cost per square centimeter of silicon detectors in the early 1990s was very high, the plan was to use silicon detectors close to the interaction point, and then to use low-cost MSGCs further away. In a space of 20 years, silicon microstrip detectors have gone from covering areas of tens of square centimeters to hundreds of square meters.

Road to assembly – silicon microstrip detectors

The detailed design of each of the subdetector units took several years, including extensive testing of prototype sensors and modules (consisting of one or two sensors and electronics), as well as the readout, cooling and power systems. After further testing and thermal cycling, modules were mounted onto low-mass carbon-fiber sub-structures with pre-assembled cooling circuits.

This project, just as many others in CMS, became a massive international logistical operation, involving work in industry as well as in CMS institutes based around the globe. In particular, CMS has pioneered the use of *automated assembly*, thus allowing for the timely construction of thousands of modules, required to deliver the 15,200 units on time. Just the requirements for circuit interconnection represented a logistical challenge, as each module was assembled from one or two microstrip sensors that had to be connected together to the readout chip. By intensive use of automatic wire bonders, these units were manufactured on time and with few delays, despite occasional variations in bond quality and rejection of sub-optimal modules.

New facilities were constructed in many institutions within the collaboration for assembling the sub-detectors as well as expanding and utilizing large laboratories. On the main CERN site, CMS built a facility (a 350-m² class-100,000 clean room) to assemble the final detector, and to provide an environment where a substantial fraction of the detection elements could be fully commissioned before installation into CMS.

As each sub-detector was assembled, it was re-tested to ensure that it continued to achieve the required performance. The integration facility was provided with rack-mounted electronics, cooling, and air-conditioning infrastructure that allowed the tracker to observe cosmic-ray events before underground installation.

Between March and August 2007 all aspects of the tracker – including safety, control and monitoring systems – were tested. Several million cosmic ray muon events were recorded at five operating temperatures between –15 °C and +15 °C. These data were reconstructed using the CMS distributed computing grid and analyzed throughout the tracker community. During this five-month period all systems operated reliably and verified that the assembled detector fully met performance specifications.

Cosmic ray data have been further analyzed, and the performance was found to be excellent. The number of inactive strips is below one part in 2000; noisy strips do not exceed 0.5%. Tracking efficiency was measured to be close to 99.8%. All of these results meet or exceed expectations and bode well for LHC physics.

The final installation of the tracker

At the CMS experimental area at Point 5 on the LHC, preparation for tracker installation started even before the CMS solenoid magnet was lowered into the cavern in February 2007. Cooling plants, power systems and off-detector readout electronics, control and data acquisition systems were all installed and tested during 2007.

18. Insertion of the tracker.

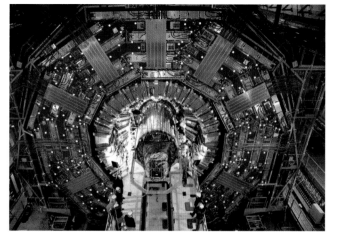

The massive task of installing cooling systems, the pipes for 450 cooling loops, 2,300 power supply cables and 400 fiber-optic cables was completed and thoroughly checked by late September 2007 (Fig. 13). The microstrip tracker was transported overnight to Point 5 on 12 December 2007, and was completely installed into CMS within two days (Fig. 18).

The connection of the services and the commissioning the tracker with the rest of CMS was completed in the Spring 2008. No degradation was noted in the good performance observed in the surface integration facility.

Lead-tungstate scintillating crystal electromagnetic calorimeter

The inner tracker was surrounded by the next layer of detection – the electro-magnetic calorimeter. CMS selected lead tungstate scintillating crystals ($PbWO_4$) as the material for this function; when a high-energy photon or electron strikes one of these crystals, the incident particle gives up its energy to the crystal by creating a shower of secondary photons, electrons, and positrons. The electrons and positrons excite the atoms, which emit *scintillation light* upon de-excitation, and this light is detected to determine the energy of the original particle. For this measurement, lead tungstate offers several advantages. It is compact (a length of 23 cm is sufficient to absorb the energy of high-energy electrons and photons); it is tolerant to high levels of radiation; and its response is fast (90% of the scintillation light is emitted within 10 nanoseconds). But it also has a number of drawbacks. The light yield is low and highly temperature dependent; this low yield necessitates the use of pho-todevices that can provide *gain* in high magnetic field. Thus, this ruled out photomultipliers for the barrel region, where *silicon avalanche photodiodes* (APDs) were eventually chosen because they could provide sufficient gain in magnetic fields.

In 1994, tests of a small number of lead tungstate crystals coupled to APDs yielded very encouraging results (see below). There remained, however, the "small matter" of producing 75,000 lead tungstate crystals, 130,000 APDs for detecting light from the barrel crystals, and 16,000 vacuum phototriodes for the endcap crystals.

172

Panel 2

Changing Economies

The procurement in Russia of lead tungstate crystals represented the largest single contract in CMS, worth over 50 million Swiss francs. A major fraction of the order was negotiated in the second half of the 1990s, when the economic conditions in Russia were difficult, and so the contracts were negotiated in US dollars. Energy represented an important part of the total cost; other important expenses included raw materials and platinum (used for lining the crucibles in which the crystals were grown). At first, the unit cost of energy in Russia was low compared to the cost in West. Gradually, this and the other costs increased, particularly when Russia decided that it would like to join the World Trade Organisation. Over this same period, the Russian currency stabilized with respect to Western currencies, leading to an inevitable increase in the US-dollar cost of producing the crystals. Indeed, the economic conditions changed so dramatically in Russia that the orders for the last 25% of the crystals were placed in Russian rubles (even though the manufacturer was given a choice of Western currencies as well).

The CMS project, during its long period of construction, faced several financial challenges of this sort; in spite of these, we were able to keep costs under control and to maintain schedules. This story of shifting international economies highlights the importance of close collaboration with industry, which was vitally important in understanding how costs could be controlled without compromising instrument performance.

From prototype to calorimeter – striving for scintillating crystals

Between 1993 and 1998, much research-and-development work was carried out to improve the transparency and the radiation hardness of the crystals. The relative amounts of lead oxide and tungsten oxide in the crystals had to be optimized; an affordable means of insuring adequate purity of the raw chemicals had to be set up; and any remaining defects in the crystals structure had to be compensated by introducing specific *dopants*. Later tests in particle beams revealed the need for a powerful laser-light monitoring system to precisely track changes in the transparency, and hence the performance, of crystals due to radiation damage. Furthermore, it was found that the crystals show recovery of transparency when the irradiation ceased, revealing a dynamic behavior that has to be carefully monitored. The mass manufacture of crystals took place between 1998 and 2008. The crystals were manufactured largely in Russia with contributions from China. In Russia a high-volume crystal growing and processing capacity (Fig. 19b) had to be put in place in a factory previously used by the military-industrial sector. The last crystals were delivered in March 2008, and the full electromagnetic calorimeter was installed by mid-August 2008.

The crystals were inserted into specially developed, lightweight carbon-fiber mechanical structures and integrated with high performance electronics. Lastly, a cooling system had to be developed that could maintain a constant temperature of approximately 100 tons of the crystals to within 0.1°C. Figure 19(c) shows the cut and polished crystals before installation in July 2007, and Figure 19(a) shows the front of an endcap 'Dee' with 5 × 5 crystals mounted in 'supercrystals'.

The electromagnetic calorimeter was commissioned for physics using penetrating muons from cosmic rays and electron "test" beams. All the indications are that the electromagnetic calorimeter is performing to specification in terms of energy resolution (approximately 0.5% at 100 GeV), noise (an energy equivalent of approximately 40 MeV per crystal), and the number of operating channels (above 99%).

19. Some views of the lead-tunstate-crystal technology used for the electromagnetic calorimeter: (a) a view of one of four "Dees" with around 3,700 endcap crystals mounted on a backplate; (b) a lead tungstate crystal ingot at the end of a 3-day growing cycle; and (c) lead tunstate crystals under inspection at CERN (photo: Peter Ginter).

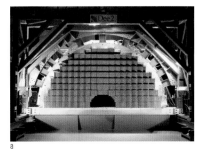

a

b

The Experiments

c

The hadron calorimeter (HCAL)

As discussed in the previous section and as seen in Figure 9, the hadron calorimeter is located between the electromagnetic calorimeter and the inner extent of the magnet coil. The hadron calorimeter is a sampling calorimeter, measuring the incident particle energy using repeated alternate layers of passive brass *absorber*, and sandwiching fluorescent scintillator plates in which special wavelength shifting optical fibers are embedded. The incoming hadron interacts essentially with brass nuclei, generating secondary particles, which in turn interact with brass nuclei to set up a shower of particles. The charged particles excite the scintillator atoms, which then emit scintillation light when they de-excite. The fibers convert the blue scintillation light to a longer wavelength which is then guided via clear optical fibers to novel hybrid photo-detectors that can provide gain when placed with their axes parallel to the field lines.

The barrel HCAL consists of two cylinders, each formed out of 18 'wedges' (Fig. 20). Each barrel-wedge module was assembled from staggered individual brass, and inner and outer stainless steel flat plates that are bolted together into a complete unit. The bolted design provided gaps into which the scintillator plates were later inserted. The novel feature is the very small clearance between the wedges (about 1 mm) to keep gaps to a minimum, through which particles might otherwise pass undetected.

The forward hadron calorimeter extends the geometric coverage to very small angles with respect to the beam ($0.8° < |\theta| < 5.7°$). A particular challenge for this detector is the unprecedented and high particle fluxes. On average about 1 TeV per proton-proton interaction is deposited into the two forward calorimeters, compared to only 100 GeV in the rest of the detector. This leads to high radiation levels, about 10 MGy over about 10 years of LHC operation at $|\theta| = 5.7°$. The design of the forward hadron calorimeter is first and foremost guided by the necessity to survive in these harsh conditions. This is the principal reason for the choice of quartz fibers as the active medium. The principle of operation is the following: *Cherenkov light* is generated when charged particles, with energy above the Cherenkov threshold, traverse the fibers (discussed in more detail in Chap. 5.5). The absorber is made by fusing 5-mm-thick steel plates with 1-mm grooves for the fibers, forming a square grid of a side of 5 mm. Figure 5.21 illustrates the insertion of quartz fibres into the absorber structure of the forward calorimeter.

20. One half of the barrel HCAL in position for insertion into the magnet vacuum tank.

21. The insertion of quartz fibers in the absorber of the forward hadron calorimeter.

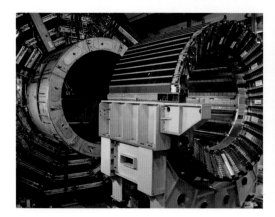

Panel 3

From Artillery to Absorbers

Russia and other member states working with the Dubna Laboratory, near Moscow, participated in the construction of the end-cap hadron calorimeter. The absorber material is brass, and it must be of sufficient quality so as to withstand high levels of stress. Concerned about cost, one of the Russian engineers remembered that artillery shell casings had the required properties. In Russian military storage, there were thousands of shells made of high-quality brass that would fit the stringent requirements – all 50 years old, made for the Navy, and designed to withstand the stresses associated with sea storage aboard 1940-era vessels. It was decided that the shell casings (Fig. 22a) would be melted for use in the CMS calorimeter.

The shells were first disarmed in a plant in northern Russia. The casings were then melted and rolled into plates in St. Petersburg, (Fig. 22b). In all, over one million WW2-vintage brass shell casings were melted down. In Minsk, Belarus, the brass plates were machined into absorber plates and pre-assembled into end-cap parts, each weighing 300 tons. These pieces were then dismounted, sent to CERN and re-mounted to form the absorber for the end caps (Fig. 22c). This project and others, including the manufacturing of the lead tungstate crystals, were part of an international agreement to convert Russian military infrastructure and know-how into peaceful technology – from swords to scientific instruments.

22. (a) Workers in Murmansk sitting on a pile of casings of decommissioned shells from the Russian Northern Fleet; (b) the rolled brass plates; and (c) the finished brass plates mounted to form the absorber of the end-cap HCAL (photo: (a) and (b) Peter Ginter).

a

b

c

175

The Experiments

The muon system

The largest part of the volume of CMS is taken up by the muon-detection system. Due to the shape of the solenoid magnet, the muon system is naturally divided into a cylindrical barrel section, completed by two planar end-cap regions. Because the muon system is comprised of approximately 25,000 m² of detection area and is positioned in relatively difficult-to-access places, the technological choices made for the muon chambers had to be inexpensive, reliable and robust.

23. The lowering of the first endcap disk showing the endcap muon CSCs.

Three types of gaseous particle detectors

In the barrel region, the rate of production of muons will be relatively low. With the uniform 4-tesla field, *drift chambers* were selected and equipped with standard rectangular drift cells. These function according to the following principle: fast charged particles, such as muons, traversing the gas inside a chamber ionize the atoms of the gas. Electron-ion pairs are created, and the electrons drift towards a thin anode wire. Close to the wire, the electric field is sufficiently high for the electrons to gain enough energy to ionize additional atoms. Very close to the wire, an exponential increase in the number of electron-ion pairs takes place, generating the signal that is picked off the anode wires. The Ar-CO_2 mixture, at atmospheric pressure, and the drift cell optics, provide a relatively linear relationship between the time elapsed after traversal and the drift path length. A point position resolution of around 300 μm can be attained in each of the eight layers in any one station, yielding the desired accuracy of 100 μm. The signal induced by the drifting charges is picked up in the anode wires.

In the end-cap regions, where the muon rate and background levels are high, and where the magnetic field is large and non-uniform, *cathode strip chambers* (CSC) are used (Fig. 23). Each six-layer CSC provides robust pattern recognition for rejection of non-muon backgrounds and efficient matching of hits to those in other stations and to the CMS inner tracker. Briefly, a particle is detected by the ionization it causes in the gas of the muon chambers during its passage. The freed electrons drift towards a thin anode wire, close to which an avalanche of further ionization electrons is created, resulting in the signal picked up by the electrodes.

In order to provide redundancy of measurement, especially for the trigger where precise timing is important, a complementary trigger system consisting of *resistive plate chambers* (RPC) is added to both the barrel and end-cap regions. The RPCs are double-gap chambers, operated in avalanche mode to ensure good operation at high rates. They produce a fast response, with good time resolution but coarser position resolution than the drift chambers or cathode strip chambers.

Event selection and acquisition

The trigger and data acquisition system usually consists of four parts, namely:
- the detector electronics;
- Level-1 trigger processors (the calorimeter, muon and global triggers);
- the readout network; and
- an on-line event-filter system (processor farm).

The LHC provides proton-proton and heavy-ion collisions, the former at high interaction rates. For protons, the beam-crossing interval is 25 ns, corresponding to a crossing frequency of 40 MHz. Since it is impossible to store and process the large amount of data associated with the resulting high number of events, a drastic selection among the events has to be achieved. This task is performed by the trigger system, which is the start of the physics event-selection process. The rate is reduced by two steps, referred to as the Level-1 Trigger (L1T), which uses custom-designed ASIC processors, and High-Level Trigger (HLT), built with commodity computer processors.

---

24. The schematic of the tracker electronics chain also showing a silicon module.

Panel 4

An Example Electronics Chain of CMS

All the electronics inside the experiment cavern have to be *radiation-tolerant*, whereas those in the high radiation environment (inner tracker, ECAL, etc.) have to be *radiation-hard*. This meant that an unusually large fraction of the electronics had to be custom-designed – a new challenge in high-energy physics. As the minimum feature size for electronics has become smaller, the cost per chip – and hence the cost per channel – has decreased. In the last decade considerable effort has gone into designing electronic components that are high performance, low-power consuming, low cost per channel, and radiation hard (Panel 5). Initially the electronics designers worked with foundries that traditionally supplied the radiation-hard electronics for military and space applications. Unfortunately, the up-front costs were high, and the long turn-around times hampered rapid development.

To illustrate the importance of "in-house" development, consider the electronics chain of the CMS tracker (Fig. 24). Each microstrip of the tracker is read out by a charge sensitive amplifier with a time constant of approximately 50 ns. The output voltage is sampled at the beam-crossing rate of 40 MHz, and the samples are stored in an analog pipeline for a duration corresponding to the Level-1 trigger latency (3.2 μs). Following a trigger, a weighted sum of three samples is formed in an analog circuit, thus confining the

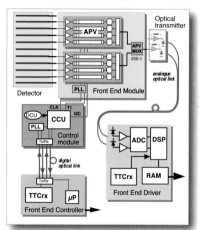

signal to a single bunch crossing and yielding the pulse height. The buffered pulse height data are multiplexed out on optical fibers, and the pulse height in each strip is used to modulate the output of a laser. The laser-light is transported via 120-m-long optical fibers to the underground control room, where the light signal is converted into an electrical signal by silicon photodiodes and digitized to give the pulse height. After *zero-suppression* and formatting, the data are stored in local memories ready for access by the data-acquisition system. Unlike most previous experiments, all of these functions are carried out by custom-designed electronics and most of the data will be transported out on optical fibers.

177

The Experiments

The rate-reduction is designed to reduce the number of recorded events by at least a factor of one million, with the combined filtering of the L1T and the HLT. The L1T, with its output rate of 100 kHz, bases its event rejection on coarse granularity data (consisting of combined channels from the calorimeters and the muon system), all the while holding full granularity and high-resolution data in pipelined memories in the front-end electronics. For those events that pass the L1T, the HLT has access to the complete read-out data and can therefore perform complex calculations similar to those made in the analysis by off-line software.

The L1T decision is based on the presence of local "trigger objects" such as photons, electrons, muons, and jets with high transverse energy or momentum. Reduced-granularity and reduced-resolution data are used to identify these trigger objects. For example, in the CMS electromagnetic calorimeter, information from groups of 25 crystals is combined to form a single "trigger-tower", for which 8-bit resolution is used instead of the full 12-bits of information. With a suitable set of transverse energy or transverse momentum thresholds, a very high acceptance rate for interesting physics can be attained while rejecting a very large fraction of uninteresting events.

During the L1T decision-making period, all the high-resolution data is held in pipelined memories. Commodity computer processors make subsequent decisions at the higher-trigger level (HLT), using more detailed information from all of the detectors. Already at this stage, the information is analyzed by sophisticated algorithms that are almost as powerful as those used for the final event reconstruction.

Panel 5
A Mid-course Correction – and a Revolution in Radiation-hard Electronics
The rapid evolution of the commercial electronics market has had a major impact on the design of CMS electronics.
In 1997, the tracker front-end chip had been almost finalized; it was found to satisfy the radiation-hardness specification and ready to go into production. The vendor, however, decided to produce the chip in a different foundry than the one used for its development.
It was soon established that radiation hardness could no longer be guaranteed.
To respond to this *force-majeure,* certain members of CMS investigated a modern, mass-production 0.25 micron CMOS technology, with the idea that the deep sub-micron processes involved would be intrinsically more resistant to radiation. However, this was found to be true only after a few subtle layout features (developed in CERN and at the Rutherford Laboratory, U.K.) were incorporated into the transistor designs. Many advantages flowed from the use of a high-volume commercial process, namely lower cost, lower intrinsic noise, lower power consumption and faster turnaround between design iterations; so much so that this class of electronics is now the obvious choice even when radiation hardness is not a requirement. Riding this technology wave became the key to managing the significant risk associated with the cost of procuring radiation-hard electronics for the CMS detector.

HLT benefits from evolution in computing technologies
In the sixteen years since the CMS Letter of Intent (1992), the speed of central processing units (CPUs) has increased by two orders of magnitude (and they now are available with eight-cores per PC "box"); memory chips hold two orders of magnitude more information; and network speeds have grown by more than a factor of 1000 – all this for a constant cost over the 16 years! The implications of this have been far reaching. It is now possible to make available the full event data, from events selected by the L1 trigger system, to a large farm of computer processors through high-bandwidth commercial switches.

Typically upon acceptance from an L1 trigger, after a fixed time interval of about 3 μs, the data from the pipelines are transferred to front-end readout buffers. After further signal processing, zero-suppression and/or data-compression, the data are placed in dual-port memories for access by the data-acquisition system. Each event, with a size of about 1 MB, is contained in several hundreds of front-end readout buffers. The data from a given event are transferred to a single commercial computer processor. Each processor runs a high-level trigger software code that reduces the Level-1 output rate of 100 kHz to a final value of 100 Hz for mass storage.

The use of a processor farm for all selections beyond Level-1 takes full advantage of the evolution in computing technology. Flexibility is maximized because there are no built-in limitations in the architecture or in the design of the system; there is complete freedom in data access and selection; and software algorithms can be updated and improved entirely free of data acquisition concerns.

Commissioning of the CMS detector

Over the course of the year 2007, the *in situ* commissioning of the detector took place in stages as the various elements were lowered into the site. Then, in a month-long round-the-clock data-taking run prior to first LHC beams in September 2008, the entire "startup" detector (Fig. 25 a,b) was commissioned. Data were taken using cosmic-ray muons to understand the performance of the installed detector elements.

CMS was thus ready to record data when the LHC started circulating beams in early September 2008. During the 40 hours of beam, we demonstrated that we could adapt our trigger and data-taking conditions to rapidly varying external conditions without losing efficiency. We were further able to show that the inter-detector time synchronization, established by the many months of running with cosmic ray triggers, worked according to plan (both at the data pipelines and at the trigger primitive generation). In addition, we found that we could further refine this synchronization using the "splashes" of several hundred thousand particles impinging all over the surface of CMS when a beam was dumped onto the last set of upstream collimators. The precision of the synchronization of the calorimeter cells was less than 2 ns. The muons in the beam halo from the first captured LHC beam (which lasted nine minutes) allowed an extraction of the alignment constants for the forward CSC muon detectors. A precision of 270 μm was achieved, which compares well with the 210 μm precision from *photogrammetry*.

Starting in mid-October 2008, a continuous month-long round-the-clock exercise was run called the Cosmic Run at Four Tesla (CRAFT), with all of CMS operational and the magnetic field at its nominal value (3.8 T). This exercise provided invaluable operational experience, permitting us to identify sources of inefficiency and to test our calibration and alignment workflows

a

b

25. (a) The CMS detector after the installation of the beam pipe in mid-2008. (b) The CMS detector was closed on 3 September 2008 in anticipation of LHC beam.

during the recording of data. We met our goal of collecting 300 million cosmic triggers with the field on, and this data provided a treasure-trove from which we were able to assess the performance of the detectors in detail. Of this number of events recorded, approximately 75 million showed muons crossing the inner strip tracker, and 75,000 of these crossed the inner pixel detector. This allowed for the first detailed extraction of alignment constants from the installed silicon tracking systems with the field on. The pixel barrel system could be aligned to a precision of about 15 μm, and the 16,000 strip tracker modules could be aligned to a precision of about 10-30 μm. The response of the calorimeter systems to muons was also measured to fine precision, with excellent agreement to theoretical predictions for energy loss, and all this took place without a single LHC collision.

Seeing new physics

CMS is not just a bigger version of recently commissioned detectors elsewhere in the world – it is a substantially different detector, thanks to the innovative and high-tech solutions found to overcome the particularly harsh environment generated by the unprecedented high luminosity of the LHC. The modular construction of CMS has brought high-energy particle detection to a new level, with the application of industrial-scale production to the building of the individual sub-detectors. The CMS detector was commissioned, and all signs indicate that it behaves as promised by the design specifications and is ready for proton-proton collisions at high energies. CMS should be capable of discovering whatever Nature has in store at the TeV energy scale and is likely to provide answers to some of the most important open questions in physics.

Acknowledgements

The construction of the large and complex experiment such as CMS could not have been carried out without the effort and the dedication of thousands of scientists, engineers and technicians worldwide. By the time the first suite of results are published many of them will have spent a substantial fraction of their professional lives on the CMS experiment. This paper is dedicated to the efforts of all these people and the agencies that funded CMS.

The Experiments

References

[1] CMS TDRs and references therein, http://cmsdoc.cern.ch/outreach/

[2] S. Chatrchyan et al., "The CMS experiment at the CERN LHC," *JINST* **3** (2008) S08004.

5.3

THE EXPERIMENTS

ATLAS

Peter Jenni

With its unprecedented collision energy and luminosity, the Large Hadron Collider opens a new frontier in particle physics. Our team has worked from the beginning to optimize the ATLAS experiment to best exploit this new territory in physics. As the reader has learned in earlier chapters, and in Chapter 2 in particular, the goal has been to maximize the discovery potential for new physics such as the discovery of Higgs bosons, supersymmetric particles and extra dimensions, while maintaining the capability to measure known objects with high accuracy, such as heavy quarks and gauge bosons. In other words, ATLAS has developed a general-purpose detector to fully cover the rich physics potential of the LHC.

Historical overview and collaboration

The ATLAS Collaboration was born of the merging of two so-called proto-collaborations, both of which presented detector concepts based on a *toroidal* muon-magnet configuration at the famous CERN – ECFA workshop 'Towards the LHC Experimental Programme' held in March 1992. During the summer of 1992 the new ATLAS Collaboration prepared its Letter of Intent (LoI) for a general-purpose proton-proton experiment at the LHC, which was submitted in October 1992 to CERN's newly formed LHC Experiments Committee (LHCC). This LoI contained a number of conceptual and technical design options that still needed to be narrowed down over the course of the following years; including the critical choice of the superconducting toroid magnet system for ATLAS. The detector concept was basically settled by the time of the submission of the ATLAS Technical Proposal (TP) to the LHCC in December 1994. The project was approved in January 1996, and the budget for the full construction was established with an expenditure ceiling set at 475 MCHF (1995 currency rate) in 1998.

ATLAS is an international collaboration, spanning the whole globe. As of April 2009, there were about 2,500 scientists working on the project as full authors, including some 700 graduate students and additional technical and administrative staff. The ATLAS Collaboration consists of 169 institutions from 37 countries, where the term "Institution" sometimes groups together several universities or research institutions of a given country. Figure 26 shows the country map of all ATLAS Collaboration Institutions, and the full list is provided in annex at the end of this chapter.

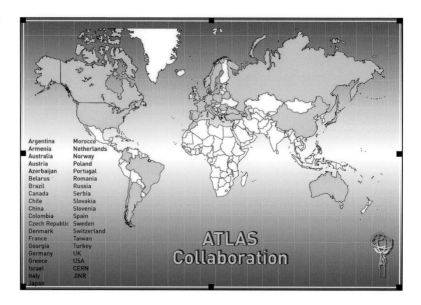

Argentina
Armenia
Australia
Austria
Azerbaijan
Belarus
Brazil
Canada
Chile
China
Colombia
Czech Republic
Denmark
France
Georgia
Germany
Greece
Israel
Italy
Japan

Morocco
Netherlands
Norway
Poland
Portugal
Romania
Russia
Serbia
Slovakia
Slovenia
Spain
Sweden
Switzerland
Taiwan
Turkey
UK
USA
CERN
JINR

**ATLAS
Collaboration**

Main design criteria

A broad spectrum of detailed physics studies drove the concept and design of the detector as described in the Technical Proposal at the end of 1994; this multi-faceted scientific instrument is now installed and ready for operation at Point 1 of the LHC. ATLAS is a versatile high-energy particle detector, with the capability to discover new – and even unanticipated – physical phenomena. Of particular interest will be those collisions that produce energetic particles emerging roughly perpendicular to the axis of the colliding beams, the so-called *high transverse momentum* (high-p_T) phenomena. To summarize the basic design criteria, ATLAS required

- very good *electromagnetic calorimetry* for electron and photon identification and measurements, complemented by full-coverage *hadronic calorimetry* for accurate jet and *missing transverse energy* (E_{Tmiss}) measurements;
- high-precision *muon momentum* measurements, with the capability to guarantee accuracy at the highest luminosity using the external muon spectrometer alone;
- efficient tracking at high luminosity for high-p_T lepton-momentum measurements, electron and photon identification, tau-lepton and heavy-flavour identification, and full-event reconstruction capability at lower luminosity;
- detection capability in almost all directions, especially along the beam axis; to express this capability, we say the detector has large acceptance in *pseudorapidity* (η), with almost full *azimuthal angle* (ϕ) coverage everywhere. The azimuthal angle is measured around the beam axis, and the pseudorapidity relates to the polar angle (θ) measured from the direction of the beam axis (described below).
- triggering and measurements of particles at low-p_T thresholds, providing high efficiencies for most physics processes of interest at the LHC;
- fast, radiation-hard electronics and sensor elements for the detectors due to the experimental conditions at the LHC, and *high detector granularity* to handle the particle fluxes and to reduce the influence of overlapping events.

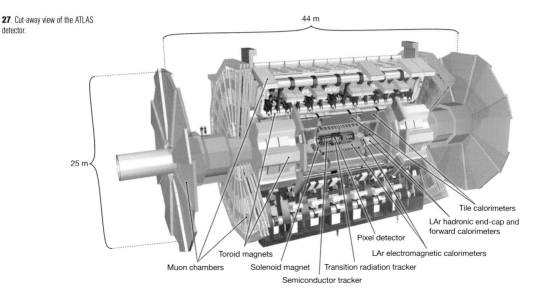

44 m

25 m

Tile calorimeters

LAr hadronic end-cap and forward calorimeters

Pixel detector

LAr electromagnetic calorimeters

Toroid magnets

Muon chambers

Solenoid magnet

Transition radiation tracker

Semiconductor tracker

Overview of the main detector components

The overall ATLAS detector layout is shown in Figure 27. A very comprehensive description, including considerable technical details, is provided elsewhere [1] in a document that includes a demonstration of its performance based on many years of test-beam measurements with real detector components.

The ATLAS detector is forward-backward symmetric with respect to the interaction point. The magnet configuration consists of a thin superconducting solenoid surrounding the inner-detector cavity, with three large superconducting toroids (one barrel and two end-caps) arranged with an eight-fold azimuthal symmetry around the calorimeters. This fundamental choice has driven the design of the rest of the detector. The powerful magnets bend the trajectories of the charged particles, and the measurement of these trajectories allows one to determine their momenta. The detector is constructed along the basic "onion skin" concept described in Chapter 5.1, and we will briefly present the role and function of each of the detection layers.

The inner detector is immersed in a 2-T solenoidal magnetic field. Its difficult task is to give a 'picture' of the hundreds of charged particles emerging from the collisions. Pattern recognition, momentum and *vertex* measurements, and electron identification are achieved with a combination of discrete, high-resolution semiconductor pixel and strip detectors in the inner part of the tracking volume; the outer part is equipped with *straw-tube tracking detectors*, which can generate and detect *transition radiation* in its outer part.

Calorimeters located as the next layer of the detector absorb most of the particles produced in the LHC interactions and measure their energy and geometrical impact position. High granularity liquid-argon (LAr) electromagnetic sampling calorimeters, with excellent performance in terms of energy and position resolution, cover the pseudorapidity range $|\eta| < 3.2$ (i.e., a coverage down to an angle of about 5° from the beam axis). The LAr technology requires the use of cryostats, as the argon has to be cooled down to temperatures of about 85 K. The hadronic calorimetry in the range $|\eta| < 1.7$ (coverage down to about 20° from the beam axis) is provided by a scintillator-tile calorimeter, which is separated into a large barrel and two smaller extended barrel cylinders, one on either side of the central barrel. In the end-caps ($|\eta| > 1.5$), LAr technology is also used for the hadronic calorimeters, matching the outer $|\eta|$ limits of end-cap electromagnetic calorimeters. The LAr forward calorimeters provide both electromagnetic and hadronic energy measurements, and extend the pseudorapidity coverage to $|\eta| = 4.9$ (i.e., as close as 1° from the axis of the beam).

The calorimeter is surrounded by the muon spectrometer designed to measure muon trajectories. The muons are the only charged particles that can penetrate the massive calorimeter layer. The air-core toroid system, with a long barrel and two inserted end-cap magnets, generates strong bending power in a large volume within a light and open structure. *Multiple-scattering effects* (the deviation of particles from their trajectory due to interaction with the material of the detector) are thereby minimized, and excellent muon momentum resolution is achieved with three layers of high precision tracking chambers. The muon instrumentation includes trigger chambers with timing resolution of the order of 1.5-4 ns. The muon spectrometer defines the overall dimensions of the ATLAS detector.

A series of detectors cover the forward regions on both sides of the ATLAS detector very near the beam line, with the aim of measuring and monitoring the LHC luminosity, as well as providing physics measurements in the very forward regions. For the start-up they are only partially installed: at 17 m from the interaction point, we have installed luminosity-monitor detectors based on *Cherenkov tubes* (LUCID); a zero-degree calorimeter for detecting photons and neutrons (mainly for heavy-ion collisions) is located at 140 m; and at 240 m, precision tracking detectors in *Roman Pots* will measure elastic scattering at very small angles for a total cross-section determination.

The proton-proton interaction rate at the design luminosity of 10^{34} cm^{-2}s^{-1} is approximately 1 GHz (1 billion collisions per second), while the event-data recording, based on technology and resource limitations, is limited to about 200 Hz (200 collisions per second). This requires an overall rejection factor of 5×10^6 against *minimum-bias processes* (low-transverse-energy interactions with high probability of production that can be mostly discarded) while maintaining maximum efficiency for the new physics. The Level-1 (L1) trigger system uses a subset of the total detector information to make a decision on whether or not to continue processing an event, reducing the data rate to approximately 75 kHz (limited by the bandwidth of the readout system, which is upgradeable to 100 kHz). The subsequent two levels, collectively known as the high-level trigger (HLT), are the Level-2 (L2) trigger and the event filter. They provide a filtering of the data to a final recording rate of approximately 200 Hz.

The magnet system

A drawing of the ATLAS superconducting magnet system can be seen in Figure 28a. It is an arrangement of a central solenoid (CS, not visible in the Figure) providing the inner detector with magnetic field, surrounded by a system of three large air-core toroids generating the magnetic field for the muon spectrometer. Overall, the magnet system is about 26 m in length and 22 m in diameter. The two *end-cap toroids* (ECT) are inserted in the giant *barrel toroid* (BT) at each end and are aligned with the central solenoid. As mentioned in Chapter 5.1, the barrel toroid determines the overall dimensions of the entire experiment. The ECT have a length of 5 m, an outer diameter of 10.7 m and an inner bore of 1.65 m. The CS extends over a length of 5.8 m and has a bore of 2.4 m. The unusual configuration and large size made the magnet system a considerable challenge requiring careful engineering.

28. (a) Schematic view of the toroid magnet system including services connections;
(b) insertion of the central solenoid into the barrel cryostat with the LAr calorimeter; and
(c) the barrel toroid installed in the underground cavern, with the barrel calorimeters including the central solenoid in the common cryostat, with the LAr in the background before its insertion into the centre of the detector.

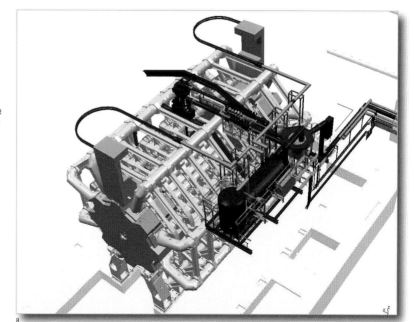

a

b

c

The central solenoid provides a central field of 2 tesla, with a peak magnetic field of 2.6 T at the superconductor itself. The peak magnetic fields on the superconductors in the barrel and endcap toroids are 3.9 and 4.1 Tesla, respectively. The performance in terms of bending power is characterized by the field integral $\int B \times dl$, where B is the azimuthal field component, and the integral is taken over the straight-line trajectory that the particle would have taken in the absence of the field, from the inner to the outer radius of the toroids; the appropriate unit for this entity is thus tesla-meters (Tm). The barrel toroid provides 2 to 6 Tm over the pseudorapidity range of 0.0-1.4; and the endcap toroid contributes with 1 to 8 Tm in the 1.6-2.7 pseudorapidity range.

When deciding to place the central solenoid in front of the electromagnetic calorimeter, careful minimization of material was required in order to achieve the desired calorimeter performance. The elegant solution was found of placing the central solenoid (CS) and the LAr calorimeters into a single common vacuum vessel, thereby eliminating two vacuum walls. The picture in Figure 28b shows how the CS is inserted into the LAr calorimeter barrel cryostat. The magnetic field of the CS has been mapped *in situ* with a relative precision of better than 10^{-4} over the full tracking volume.

Each of the three toroids consists of eight coils, assembled radially and symmetrically around the beam axis. The end-cap-toroid coil system is rotated by 22.5° with respect to the barrel-toriod coil system in order to provide radial overlap and to optimize the bending power in the regions between the coil systems. The barrel-toroid coils are of a flat racetrack type with two double-pancake windings made of 20.5 kA aluminum-stabilised Nb-Ti superconductor. The windings are housed in an aluminum alloy casing, with the magnetic forces transferred to the warm structure. Each of the eight coils are housed in an individual cryostat; the toroidal structure consists therefore of these eight cryostats and their linking elements, called voussoirs and struts, that provide mechanical stability. Services are brought to the coils through a cryogenic ring linking the eight cryostats to a separate service cryostat, which provides connections to the power supply, the helium refrigerator, the vacuum systems and the control system. Figure 28c shows the barrel toroid after its installation in the underground cavern.

Each end-cap toroid (ECT) also consists of eight coils in an aluminium alloy housing. In contrast to the barrel toroid, they are cold-linked and assembled as a single cold mass, housed in one large cryostat. Therefore the internal forces in the toroids are taken up by the cold supporting structure between the coils. Due to the magnetic forces, the ECT magnets are pulled into the barrel toroid and the corresponding axial forces are transferred to the BT via axial transfer points linking both magnet systems. The cryostats rest on a rail system facilitating the pulling back and parking of the ECT magnets, thus providing access to the detector center. The magnets are indirectly cooled by a forced flow of helium at 4.5 K through tubes welded on the casing of the windings. The cooling power is supplied by a central refrigeration plant located in the side cavern and the services are distributed among the four magnets.

Electrically, the eight coils of the barrel toriod are connected in series, as are the 16 coils in the two end-cap toroids. The toroid-coil systems have a 21 kA power supply and are equipped with control systems for fast and slow energy dumps. The central solenoid is energized by an 8 kA power supply. An adequate and proven quench protection system has been designed to safely dissipate the total stored energy of 1.6 GJ without overheating the coil windings. The full magnet system, including its fast safety-discharge procedures, is now fully commissioned in its final configuration in the cavern.

The inner detector

Approximately 1000 particles will emerge from the collision point every 25 ns within $|\eta| < 2.5$, creating a very large track density in the detector. To achieve the momentum and vertex resolution required to observe the anticipated physics processes, high-precision measurements must be made along the trajectories of the particles with fine detector granularity. Pixel and silicon microstrip (SCT) semiconductor trackers, used in conjunction with the straw tubes of the transition radiation tracker (TRT), offer these features. A view of these three elements and their positioning relative to the beam axis is found earlier in this Chapter in Figure 4(a).

The layout of the inner detector (ID) is illustrated in Figure 29(a). The precision tracking detectors (pixels and silicon microstrip) cover the region $|\eta| < 2.5$. In the barrel region, they are arranged on concentric cylinders around the beam axis while in the end-cap regions they are located on disks perpendicular to the beam axis.

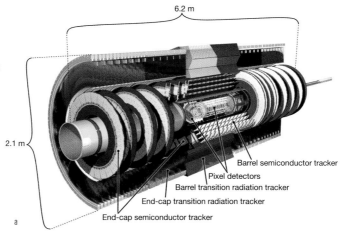

29. (a) Cut-away view of the ATLAS inner detector; (b) the barrel SCT tracker in front of the barrel TRT detector, just before the integration in the surface clean room; and (c) the insertion of the completely preassembled pixel detector into the center of ATLAS.

6.2 m

2.1 m

Barrel semiconductor tracker
Pixel detectors
Barrel transition radiation tracker
End-cap transition radiation tracker
End-cap semiconductor tracker

a

b

c

The Experiments

The highest detection granularity is achieved around the vertex region through the use of silicon pixel detectors. The pixel layers are segmented both perpendicularly to the beam axis and in the direction parallel to the beam axis; typically three pixel layers will be crossed by each track. All pixel sensors are identical and have a minimum pixel size of $50 \times 400 \ \mu m^2$, resulting in a total readout of approximately 80.4 million channels. For the microstrip detectors, eight strip layers (four space points) are crossed by each track. Located in both the barrel and end-cap regions, the microstrip detectors provide a total of about 6.3 million readout channels.

A large number of hits (typically 36 per track) along the particle trajectories are provided by the 4-mm-diameter straw tubes of the transition radiation tracker (TRT), which can follow tracks up to a psuedorapidity of $|\eta| = 2.0$. The TRT consists of hundreds of thousands of "straws"; each straw is in reality a small cylindrical gas-containing chamber, with an anode wire in the center, and the wall of the straw acts as a cathode. Charged particles and photons passing through the straw ionize the gas, producing a measurable current. In particular, when relativistic electrons interact with the gas-solid interface of material near the straws, *transition-radiation photons* are generated and subsequently detected by ionization of the TRT gas mixture.

The TRT only provides information in the azimuthal direction, for which it has an intrinsic accuracy of $130 \ \mu m$ per straw. In the barrel region, the straws run parallel to the beam axis and are 144 cm long, with their wires divided into two halves at the point that corresponds to approximately at $\eta = 0$ (perpendicular to beam line at the collision vertex). In the end-cap region, the

37-cm-long straws are radially arranged in wheels. There are approximately 351,000 TRT readout channels.

The combination of precision trackers at small radii with the TRT at a larger radius gives very robust pattern recognition and high precision in all detection directions. The signal from the TRT at the outer radius contributes significantly to the momentum measurement; even if the straw detectors have lower precision per point compared to the silicon detectors, the TRT has the advantage of providing a large number of measurements at longer track length.

The inner-detector system tracks over a range matched by the precision measurements of the electromagnetic calorimeter. The electron-identification capabilities are enhanced by the detection of transition-radiation photons in the xenon-based gas mixture of the straw tubes. The semiconductor trackers also measure the part of the track closest to the interaction point, allowing reconstruction of the possible *secondary vertices*, originated by the decays of short-lived particles – such as hadrons containing a heavy quark (b or c) or the tau lepton – into more ordinary, long-lived particles. These particles are characterized by a measurable flight path of typically 0.1 to 1 millimeters, depending on their energy, before the decay. The resulting secondary vertex, and the quality of its measurement, is enhanced by the innermost layer of pixels, located at a radius of about 5 cm from the collision point.

Figures 29(b) and (c) show pictures of the preassembly in a clean room at the surface, and of the final insertion at the center of the ATLAS detector in the underground cavern.

Calorimetry

A view of the ATLAS *sampling calorimeters* (i.e., made of a combination of passive and active materials) is presented in Figure 30(a). They cover the range $|\eta| < 4.9$ through the use of different techniques, suited to the widely varying requirements of the physics processes of interest and of the radiation environment over this large range of pseudorapidity. Over the region matched to the inner detector, the fine granularity of the electromagnetic calorimeter is used for precise measurement of electrons and photons. Outside this range, the coarser granularity of the calorimeter is sufficient to satisfy the physics requirements for jet reconstruction and measurement of missing transverse energy (as described in the first section of this Chapter).

The calorimeters must provide good containment for electromagnetic and hadronic showers, and must also limit *punch-through* (i.e., particles leaking out of the calorimeter) into the muon system. Thicker calorimeters improve the punch-through affects of the detector; however this has to be balanced against the increased weight and cost of the device. This is quantified through two parameters: the *radiation length* (X_0), which is defined as the average distance travelled by an electron before encountering an electromagnetic interaction inside the colorimeter; and the *interaction length* (λ), defined as the average distance for a hadron to undergo a hadronic interaction. The total thickness of the electromagnetic calorimeter is more than $22\ X_0$ in the barrel and exceeds $24\ X_0$ in the end-caps. The approximate 9.7 interaction lengths of active calorimeter in the barrel ($10\ \lambda$ in the end-caps) are adequate to provide good resolution for high energy jets. The total thickness, including $1.3\ \lambda$ from the outer support, is $11\ \lambda$ at $\eta = 0$; this has been shown, both by measurements and simulations, to be sufficient to reduce punch-through well below the irreducible level of prompt or decay muons.

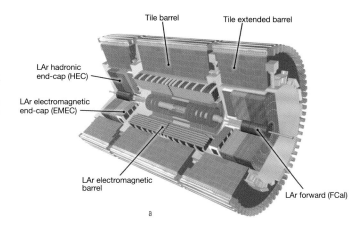

Tile barrel

Tile extended barrel

LAr hadronic
end-cap (HEC)

LAr electromagnetic
end-cap (EMEC)

LAr electromagnetic
barrel

LAr forward (FCal)

a

30. (a) Cut-away view of the ATLAS calorimetry; (b) the barrel module of the LAr electromagnetic calorimeter during stacking of absorbers and readout electrodes; and (c) the fully installed end-cap calorimeter (LAr EM, HEC and FCal in a common cryostat, surrounded by the tile calorimeter) partially inserted into the barrel toroid region.

b

c

Liquid argon (LAr) electromagnetic calorimeter

The electromagnetic calorimeter (EM) is the innermost of the calorimeters and is divided into a barrel part ($|\eta| < 1.475$) and two end-cap components ($1.375 < |\eta| < 3.2$), each housed in their own cryostat. The barrel calorimeter consists of two identical half-barrels, separated by a small gap (4 mm) at $z = 0$. Each end-cap calorimeter is mechanically divided into two coaxial wheels: an outer wheel covering the region $1.375 < |\eta| < 2.5$, and an inner wheel covering the region $2.5 < |\eta| < 3.2$. The EM calorimeter is a lead/liquid-argon detector with accordion-shaped kapton electrodes and lead absorber plates over its full coverage. When high-energy electrons or photons pass through the lead, an *electron shower* is produced, with an intensity proportional to the incident energy. Argon, as the active material, is ionized in the presence of the showers, producing a current, also proportional to the energy of the incoming particle, that can be measured. The accordion geometry provides complete azimuthal coverage without cracks (i.e., full 360° coverage around the beam axis). Figure 30(b) shows a barrel module during stacking of the absorber and electrodes. The lead thickness in the absorber plates has been optimized as a function of pseudorapidity for the energy resolution capabilities of the EM calorimeter. Over the region devoted to precision physics ($|\eta| < 2.5$), the EM calorimeter is segmented into three sections in depth with high granularity. For the end-cap inner wheel, the calorimeter is two sections in depth, and it has a coarser lateral granularity compared to the other part of the detection volume. In the region of $|\eta| < 1.8$, a *pre-sampling detector* (an LAr gap located in front of the calorimeter with a separate read-out) is used to correct for the energy lost by electrons and photons upstream of the calorimeter. The EM calorimeter has, in total, about 175,000 readout channels.

Hadronic calorimeters

The hadronic calorimeters have two main functions. First, they measure the energies and directions of *jets* (i.e., clusters of particles) that result from the ejection of quarks or gluons. These particles are not directly measurable, but they appear in the form of particle clusters emerging from the collision point. Secondly, the hadronic calorimeter is used to infer the production of one or more undetectable neutral particles by monitoring the total transverse momentum; the detection of *missing transverse energy* (E_{Tmiss}) is a strong signal that such particles were produced in a collision. There are three types of hadronic calorimeters built into ATLAS: the tile calorimeter, the hadronic end-cap calorimeter, and the forward calorimeter, each discussed below in more detail.

The *tile calorimeter* is placed directly outside the EM calorimeter envelope and consists of steel-plates as the absorber combined with scintillating tiles as the active material. Its barrel covers the region $|\eta| < 1.0$, and its two extended barrels cover the range $0.8 < |\eta| < 1.7$. The barrel and extended barrels are divided azimuthally into 64 modules. Radially, the tile calorimeter extends from an inner radius of 2.28 m to an outer radius of 4.25 m. It is segmented in depth in three layers, approximately 1.5, 4.1 and 1.8 *interaction lengths* (λ) thick for the barrel and 1.5, 2.6, and 3.3 λ for the extended barrel. Two sides of the scintillating tiles are read out by wavelength shifting fibers into two separate photomultiplier tubes, providing roughly 10,000 readout channels.

The *hadronic end-cap calorimeter* (HEC) consists of two independent wheels per end-cap, located directly behind the end-cap electromagnetic calorimeter and sharing the same LAr cryostats; however, copper is used as the absorber. To reduce the drop in material density at the transition between the end-cap and the forward calorimeter (around $|\eta| = 3.1$), the HEC extends out to $|\eta| = 3.2$, thereby overlapping with the forward calorimeter. Similarly, the HEC η range extends to $|\eta| = 1.5$ and thus also slightly overlaps that of the tile calorimeter ($|\eta| < 1.7$). Each wheel is built from 32 identical wedge-shaped modules, assembled with fixtures at the periphery and at the central bore. Each wheel is divided into two segments in depth, for a total of four layers per end-cap. The wheels closest to the interaction point are built from 25-mm parallel copper plates, while those further away use 50-mm copper plates (for all wheels the first plate is half-thickness). The outer radius of the copper plates is 2.03 m, while the inner radius is 0.475 m (except in the overlap region with the forward calorimeter where this radius becomes 0.372 m). The copper plates are interleaved with 8.5-mm liquid-argon gaps, providing the active medium for this sampling calorimeter. The granularity decreases towards larger $|\eta|$, resulting in about 5,600 readout channels.

The *forward calorimeter* (FCal) is integrated into the end-cap cryostats, as this provides clear benefits in terms of uniformity of the calorimetric coverage as well as reduced radiation background levels in the muon spectrometer. The FCal is approximately 10 interaction lengths deep, and consists of three modules in each end-cap: the first, made of copper, is optimized for electromagnetic measurements, while the other two, made of tungsten, measure predominantly the energy of hadronic interactions. Each module consists of a metal matrix, with regularly spaced longitudinal channels filled with the electrode structure consisting of concentric rods and tubes parallel to the beam axis. Liquid argon in the gap between the rod and the tube is the sensitive medium. This geometry allows for excellent control of the gaps, which are as small as 0.25 mm in the first section, in order to avoid problems due to ion buildup. Rods are grouped for the readout totaling about 3,500 channels.

The photograph Figure 30(c) shows the fully installed end-cap calorimeter partially inserted into the barrel-toroid region of the ATLAS detector.

The muon spectrometer

The conceptual layout of the muon spectrometer is shown in Figure 31. It is based on the magnetic deflection of muon tracks in the large superconducting air-core toroid magnets, instrumented with separate trigger and high-precision tracking chambers. Over the range $|\eta| < 1.4$, magnetic bending is provided by the large barrel toroid. For $1.6 < |\eta| < 2.7$, muon tracks are bent by two smaller end-cap magnets inserted into both ends of the barrel toroid. Over $1.4 < |\eta| < 1.6$, usually referred to as the transition region, magnetic deflection is provided by a combination of barrel and end-cap fields. This magnet configuration provides a field mostly perpendicular to the muon trajectories, while minimizing the degradation of resolution due to multiple scattering because, in this configuration, the muons will travel mainly through the air.

The anticipated high level of particle flux has had a major impact on the choice and design of the spectrometer instrumentation, affecting performance parameters such as rate capability, granularity, ageing properties, and radiation hardness. In the barrel region, tracks are measured in chambers arranged in three cylindrical layers around the beam axis; in the transition and end-cap regions, the chambers are installed in planes perpendicular to the beam, also in three layers.

Over most of the η range, a precision measurement of the track coordinates is made in one direction, specifically in the principal bending direction of the magnetic field, and is provided by *monitored drift tubes* (MDTs). These sensors are made of gas-filled tubes, each with a single wire running axially down the center. High voltage between the wire and the tube wall produces ionization electrons when muons pass through the tube volume, thus allowing their trajectories to be tracked. The mechanical isolation in the drift tubes of each sense wire from its neighbors guarantees a robust and reliable operation. In the final configuration the muon spectrometer will consist of about 350,000 tubes (each terminating in an individual readout channel) grouped in 1,150 chambers, currently 95% of them are installed and operational. At angles close to the beamline, 32 cathode strip chambers (CSCs, which are multiwire proportional chambers with cathodes segmented into strips) with higher granularity (31,000 readout channels) are used in the innermost plane over the range $2 < |\eta| < 2.7$, to withstand the demanding rate and background conditions.

The stringent requirements on the relative alignment of the muon chamber layers are met by the combination of precision mechanical-assembly techniques and optical alignment systems both within and among muon chambers.

The trigger system to select events with muon candidates operates for $|\eta| < 2.4$. Resistive plate chambers (RPCs) are used in the barrel and thin gap chambers (TGCs) in the end-cap regions. The trigger chambers for the muon spectrometer serve a threefold purpose, namely they provide *bunch-crossing* identification and well-defined p_T thresholds; and they measure the muon coordinate in the direction perpendicular to that determined by the precision-tracking chambers (which, as mentioned above, measures position only along

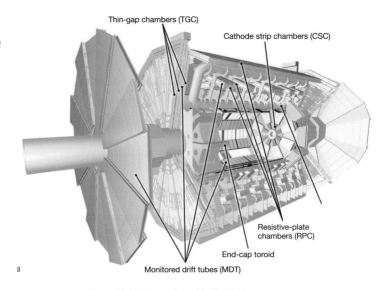

Thin-gap chambers (TGC)

Cathode strip chambers (CSC)

Resistive-plate
chambers (RPC)

End-cap toroid

a   Monitored drift tubes (MDT)

31. (a) Cut-away view of the ATLAS muon spectrometer instrumentation; and (b) the large assemblies ('Big Wheels') of the ATLAS end-cap muon chambers, TGCs on the left and MDTs on the right, with the forward shielding around the beam pipe visible in the center.

b

one direction). The total number of RPCs is about 600 (375,000 channels), and the total number of TGCs is about 3,600 (320,000 channels).

The overall performance over the large areas involved, particularly at the highest momenta, depends on the alignment of the muon chambers with respect to one another as well as to the overall detector. The accuracy of the stand-alone muon momentum measurement necessitates a precision of 30 μm on the relative alignment of chambers both within each *projective tower* and between consecutive layers in immediately adjacent towers. The internal deformations and relative positions of the MDT chambers are monitored by approximately 12,000 precision-mounted alignment sensors, all based on the optical monitoring of deviations from straight lines. Because of geometrical constraints, the reconstruction and/or monitoring of the chamber positions rely on somewhat different strategies and sensor types in the end-cap and barrel regions.

The accuracy required for the relative positioning of non-adjacent towers to obtain adequate mass resolution for multi-muon final states, lies in the few millimeter range. This initial positioning accuracy is approximately established during the installation of the chambers. Ultimately, the relative alignment of the barrel and forward regions of the muon spectrometer, of the calorimeters and of the inner detector will rely on high-momentum muon trajectories that produce almost straight tracks.

For magnetic field reconstruction, the goal is to determine the bending power along the muon trajectory to within a few parts in a thousand. The field is continuously monitored by a total of approximately 1,800 Hall sensors distributed throughout the spectrometer volume. Their readings are compared with magnetic-field simulations and used for reconstructing the position of the toroid coils in space, as well as to account for magnetic perturbations induced by the tile calorimeter and other nearby metallic structures. Figure 31(b) shows the large assemblies of the muon detectors in the end-cap regions of ATLAS.

Trigger and Data Acquisition

The Trigger and Data Acquisition (TDAQ) systems, the timing- and trigger-control logic, and the Detector Control System (DCS) are partitioned into sub-systems, typically associated with sub-detectors, which are made of the same logical components and building blocks. The trigger system has three distinct levels: L1, L2, and the event filter. Each trigger level refines the decisions made at the previous level and, where necessary, applies additional selection criteria. The data acquisition system receives and buffers the event data from the detector-specific readout electronics, at the L1 trigger accept rate, across 1,600 point-to-point readout links. The first level uses a limited amount of the total detector information to make a decision in less than 2.5 μs, reducing the rate to about 75 kHz. The two higher levels access more detector information for a final rate of up to 200 Hz with an event size of approximately 1.3 MB. A schematic view of the TDAQ system is given in Figure 32.

Trigger system

The L1 trigger searches for high transverse-momentum muons, electrons, photons, jets, and tau-leptons decaying into hadrons, as well as large missing and total transverse energy. Its selection is based on information from a subset of detectors. High transverse-momentum muons are identified using trigger chambers in the barrel and end-cap regions of the spectrometer. Calorimeter selections are based on reduced-granularity information from all the calorimeters. Results from the L1 muon and calorimeter triggers are processed by the central trigger processor, which implements a trigger 'menu' made up of combinations of trigger selections.

Events passing the L1 trigger selection are transferred to the next stages of the detector-specific electronics, and subsequently to the data acquisition via point-to-point links. In each event, the L1 trigger also defines one or more Regions-of-Interest (RoI's), i.e. the geographical coordinates of those regions within the detector where its selection process has identified interesting features. The RoI data include information on the type of feature identified, and the criteria satisfied (e.g. a particular threshold). The high-level trigger subsequently uses this information in the next stage of evaluation.

The L2 selection is thus "seeded" by the RoI information provided by the L1 trigger over a dedicated data path. L2 selections use, at full granularity and

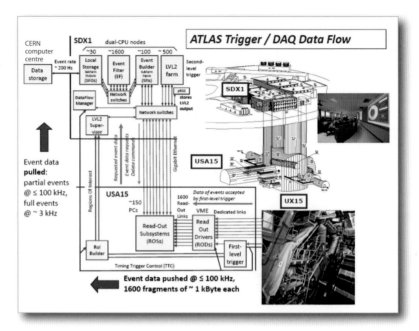

precision, all the available detector data within the RoIs (approximately 2% of the total event data). The L2 menus are designed to reduce the trigger rate to approximately 3.5 kHz, with an event processing time of about 40 ms, averaged over all events. The final stage of the event selection is carried out by the event filter, which reduces the event rate to roughly 200 Hz. Its selections are implemented using offline-analysis procedures within an average event processing time on the order of four seconds.

Readout architecture and data acquisition

The Readout Drivers (RODs) are detector-specific functional elements of the front-end electronics systems, which achieve a higher level of data concentration and multiplexing by gathering information from several front-end data streams. Although each sub-detector uses specific front-end electronics and RODs, these components are built from standardized blocks and are subject to common requirements. The front-end electronics sub-system includes different functional components:

- the front-end analogue processing or analogue-to-digital processing;
- the L1 buffer in which the (analogue or digital) information is retained for a time long enough to accommodate the L1 trigger latency;
- the de-randomizing buffer, in which the data corresponding to a L1 trigger are stored before being sent to the following level. This element is necessary to accommodate the maximum instantaneous L1 rate without introducing significant dead-time.

After an event is accepted by the L1 trigger, the data from the pipe-lines are transferred off the detector to the RODs. Digitized signals are formatted as raw data prior to being transferred to the DAQ system.

The first stage of the DAQ – the readout system – receives and temporarily stores the data in local buffers. It is subsequently solicited by the L2 trigger for the event data associated with the RoIs. Those events selected by the L2 trigger are then transferred to the event-building system and subsequently

to the event filter for final selection. Events selected by the event filter are moved to permanent storage at the CERN computer center for offline data analysis using the Worldwide LHC Computing Grid (WLCG) as the backbone (see Chap. 5.6). In addition to the movement of data, the data acquisition also provides for the configuration, control and monitoring of the hardware and software components which together provide the data-taking functionality.

The detector control system (DCS) permits the coherent and safe operation of the ATLAS detector hardware, and serves as a homogeneous interface to all sub-detectors and to the technical infrastructure of the experiment. It controls, continuously monitors, and archives the operational parameters; it signals any abnormal behavior to the operator; and it allows automatic or manual corrective actions to be taken. Typical examples are high- and low-voltage systems for detector and electronics; gas and cooling systems; magnetic field, temperatures, and humidity. The DCS also enables bi-directional communication with the data-acquisition system in order to synchronize the state of the detector with data taking. It also handles the communication between the sub-detectors and other systems independently controlled systems, such as the LHC accelerator, the CERN technical services, the ATLAS magnets, and the detector safety system.

Commissioning with cosmic rays and outlook

The ATLAS detector has been brought into operation over a two-year period as components were installed and cabled. The full installation and integration of the detector was in itself an enormous technical challenge [1], requiring a total of about five years. To give the reader an idea of the magnitude of the task, it was necessary to pull more than 50,000 cables with a total length exceeding 3,000 kilometers, and to install 10,000 pipes and tubes for servicing the detector.

Figure 33 shows the main ATLAS Control Room where hectic around-the-clock activity has already begun to commission the overall detector in view

197

The Experiments

of the imminent start of the LHC operation. It is beyond the scope of this chapter to discuss any of the preparatory work for the forthcoming LHC data analysis. Observation of the cosmic rays as they transverse the detector allows the collaboration to practice all steps, from triggering and collecting the data to the final offline analysis using the WLCG distributed computing resources in the international collaboration. Figure 34 shows an 'event display' of a cosmic ray muon registered in ATLAS and reconstructed through the full computing chain.

ATLAS is now ready to start operation at the LHC, and the large community is eagerly waiting to explore the exciting physics potential of the LHC. The ATLAS detector has been designed to operate for at least ten years at the design luminosity of the LHC. However, a vigorous research-and-development program for some detector components has already started within the Collaboration to prepare for a possible future high-luminosity upgrade phase of the LHC and the ATLAS detector which could extend the lifetime of this fantastic project for another decade.

Further reading
[1] G. Aad et al., The ATLAS Collaboration, "The ATLAS Experiment at the CERN Large Hadron Collider" *JINST* **3** (2008) S08003.

Members of the ATLAS collaboration:

Albany, Alberta, NIKHEF Amsterdam, Ankara, LAPP Annecy, Argonne NL, Arizona, UT Arlington, Athens, NTU Athens, Baku, IFAE Barcelona, Belgrade, Bergen, Berkeley LBL and UC, HU Berlin, Bern, Birmingham, UAN Bogota, Bologna, Bonn, Boston, Brandeis, Bratislava/SAS Kosice, Brookhaven NL, Buenos Aires, Bucharest, Cambridge, Carleton, Casablanca/Rabat, CERN, Chinese Cluster, Chicago, Chile, Clermont-Ferrand, Columbia, NBI Copenhagen, Cosenza,

AGH UST Cracow, IFJ PAN Cracow, UT Dallas, DESY, Dortmund, TU Dresden, JINR Dubna, Duke, Frascati, Freiburg, Geneva, Genoa, Giessen, Glasgow, Göttingen, LPSC Grenoble, Technion Haifa, Hampton, Harvard, Heidelberg, Hiroshima, Hiroshima IT, Indiana, Innsbruck, Iowa SU, Irvine UC, Istanbul Bogazici, KEK, Kobe, Kyoto, Kyoto UE, Lancaster, UN La Plata, Lecce, Lisbon LIP, Liverpool, Ljubljana, QMW London, RHBNC London, UC London, Lund, UA Madrid, Mainz, Manchester, CPPM Marseille, Massachusetts, MIT,

Melbourne, Michigan, Michigan SU, Milano, Minsk NAS, Minsk NCPHEP, Montreal, McGill Montreal, FIAN Moscow, ITEP Moscow, MEPhI Moscow, MSU Moscow, Munich LMU, MPI Munich, Nagasaki IAS, Nagoya, Naples, New Mexico, New York, Nijmegen, BINP Novosibirsk, Ohio SU, Okayama, Oklahoma, Oklahoma SU, Olomouc, Oregon, LAL Orsay, Osaka, Oslo, Oxford, Paris VI and VII, Pavia, Pennsylvania, Pisa, Pittsburgh, CAS Prague, CU Prague, TU Prague, IHEP Protvino, Regina, Ritsumeikan,

UFRJ Rio de Janeiro, Rome I, Rome II, Rome III, Rutherford Appleton Laboratory, DAPNIA Saclay, Santa Cruz UC, Sheffield, Shinshu, Siegen, Simon Fraser Burnaby, SLAC, Southern Methodist Dallas, NPI Petersburg, Stockholm, KTH Stockholm, Stony Brook, Sydney, AS Taipei, Tbilisi, Tel Aviv, Thessaloniki, Tokyo ICEPP, Tokyo MU, Toronto, TRIUMF, Tsukuba, Tufts, Udine/ICTP, Uppsala, Urbana UI, Valencia, UBC Vancouver, Victoria, Washington, Weizmann Rehovot, FH Wiener Neustadt, Wisconsin, Wuppertal, Würzburg, Yale, Yerevan.

34. Event display of a cosmic ray muon traversing the ATLAS detector.

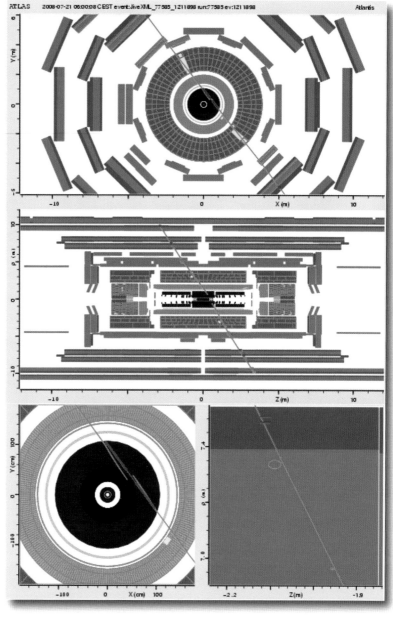

The Experiments

5.4

THE EXPERIMENTS

ALICE
Juergen Schukraft
and Christian Fabjan

ALICE, an acronym for A Large Ion Collider Experiment, is very different in both design and purpose from the other experiments at the LHC. It aims to study head-on collisions between heavy nuclei, at first mainly lead with lead at the top energy of the LHC. In these reactions, the LHC's enormous energy – collisions of lead nuclei are 100 times more energetic than those of protons – will heat up the matter to a temperature 100,000 times higher than the temperature in the core of our sun. Nuclei and nucleons melt into their elementary constituents, quarks and gluons, to form for a brief instant the primordial matter that filled the universe until a few microseconds after the Big Bang. The hot reaction zone expands at almost the speed of light; during this expansion it cools, breaks up and condenses back into a plethora of ordinary, composite matter particles.

The ALICE detector has to measure as many of the escaping particles as possible; several tens of thousands are produced in each of these 'little bangs'. The information concerning the quantity, type, mass, energy and direction of all the particles will be required in order to infer the existence and properties of matter under the extreme conditions created during the instant of the collision.

ALICE will also take data from proton-proton collisions. Comparing these with the heavy-ion results will allow us to look for tell-tale changes between the two types of beams and to characterize the global-event structure of proton reactions with its set of detectors, which are very different and complementary to the other LHC experiments.

History and Challenges

The physics program of high-energy heavy-ion collisions and the search for the *quark-gluon plasma* (QGP), the primordial matter of the universe, started in 1986 at the CERN SPS accelerator and, simultaneously, at the Brookhaven Alternating Gradient Synchrotron in the USA. The first set of detectors, many of them put together from equipment used in previous generations of experiments, could actually only use rather light ion beams (from oxygen to silicon) at what today would be considered as rather modest energy for these collisions at fixed targets (5 - 20 GeV center-of-mass energy). Already the following year, in 1987, during a workshop to choose CERN's next accelerator project, the possibility of using both heavy ions as well as protons was mentioned for the machine which was to become the LHC. In 1990, when the Relativistic Heavy Ion Collider (RHIC) – the first heavy-ion collider, built in the USA – was approved and a call for experiments issued, the European community faced the decision of either participating in the RHIC or focusing its resources on the LHC. The schedules for RHIC and LHC were, at the time, quite comparable; a sequential exploitation of both machines seemed implausible. A series of workshops and discussions were held to look at the physics potential of these machines and at different detector concepts. Then the Europeans made the decision, correct as it turns out with hindsight, to participate at a modest scale at RHIC and to start in parallel a dedicated design and research-and-development effort for a large general purpose heavy ion detector at LHC. This left the community with a busy schedule and thinly stretched resources, with commitments towards ongoing data analysis of the light-ion program; construction of a new generation of experiments for the heavy-ion program starting at CERN with lead (Pb) beams in 1994; designing and building detectors for RHIC (in operation since 2000); and further research and development for an ambitious new experiment at LHC. All this happened in parallel and involved many of the same actors and groups, but it did pay off handsomely in a rich program and fast progress.

Designing a dedicated heavy-ion experiment in the early 1990s for use at the LHC some 15 years afterwards posed some daunting challenges: in a field still in its infancy, it required extrapolating the conditions to be expected by a factor of 300 in energy and a factor of 7 in beam mass. The detector therefore had to be both "general purpose" – able to measure most signals of potential interest, even if their relevance may only become apparent later – and flexible, allowing additions and modifications along the way as new avenues of investigation would open up. In both respects ALICE did quite well, as it included a number of observables in its initial menu, the importance of which only became clear after results appeared from RHIC. Various major detection systems where added over time to match the evolving physics, from the muon spectrometer in 1995, the transition-radiation detector in 1999, to a large-jet calorimeter added as recently as 2007.

The anticipated experimental conditions for nucleus-nucleus collisions at the LHC posed perhaps the greatest challenge, with the extreme number of particles produced in every single event. Particle production was expected to be three orders of magnitude larger than in typical proton-proton interactions at the same energy, and even a factor two to five above the highest multiplicities measured at RHIC. On the other hand, the time between collisions will be four times longer in the heavy-ion configuration relative to the proton-proton configuration and the luminosity (i.e, the probability of actually having a collision) much lower; this allows the use of "slower" detection techniques compared to ATLAS and CMS. The tracking of these particles was therefore made particularly safe and robust by using mostly three-dimensional hit information with many points along each track (up to 150) in a moderate magnetic field. An overly strong field would both mix up the particles and exclude the lowest energy ones from being observed by imposing a small radius of curvature on the particle trajectory (the paths become more curved at high magnetic field and low momentum).

In addition, a large dynamic range is required for momentum measurement, spanning more than three orders of magnitude from tens of MeV to well over 100 GeV. This is achieved with a combination of detectors built with very low material thickness (to reduce scattering of low momentum particles) and a large tracking lever arm of up to 3.5 m (resolution improves at high momentum with the square of the measurement length), thus achieving good resolution at both high and low momentum with a modest field. And finally, particle identification (PID) over much of this momentum range is essential, as many phenomena depend critically on either particle mass or particle type. ALICE therefore employs essentially all known PID techniques in a single experiment, as discussed in some detail in the following sections.

The ALICE design evolved from the Expression of Interest (1992) via a Letter of Intent (1993) to the Technical Proposal (1996) and was officially approved in 1997. The first ten years were spent on design and an extensive research and development effort. As for all other LHC experiments, it became clear from the outset that also the challenges of heavy-ion physics at LHC could not be really met (nor paid for) with existing technology. Significant advances, and in some cases a technological break-through, would be required to build on the ground what physicists had dreamed up on paper for their experiments. The R&D effort started with broad goals but rapidly became more focused, organized and supported over most of the 1990s, leading to many evolutionary – and some revolutionary – advances in detectors, electronics and computing.

ALICE Detector Overview

ALICE is usually referred to as one of the small detectors, but the meaning of "small" is very relative in the context of LHC: The detector stands 16 meters tall, is 16-m wide and 26-m long, and weighs in at approximately 10,000 tons. It has been designed and built over almost two decades by a collaboration of over 1000 scientists and engineers from more than 100 Institutes in some 30 different countries. The experiment consists of 17 different detection systems, each with its own specific technology choice and design constraints.

A schematic view of ALICE is shown in Figure 35. It consists of a central part, which measures hadrons, electrons, and photons, and a forward sin-

203

The Experiments

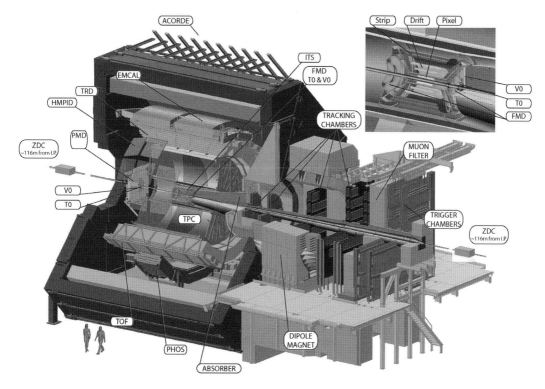

gle-arm spectrometer that focuses on muon detection. The central 'barrel' part covers the direction perpendicular to the beam from 45° to 135° and is located inside a huge solenoid magnet, which was built in the 1980s for the L3 experiment at CERN's LEP accelerator. As a warm resistive magnet, the maximum field at the nominal power of 4 MW reaches 0.5 T. The central barrel contains a set of tracking detectors, which record the momentum of the charged particles by measuring their curved path inside the magnetic field. These tracks are then identified according to mass and particle type by a set of particle identification detectors, followed by two types of electromagnetic calorimeters for photon and jet measurements. The forward muon arm (2° - 9°) consists of a complex arrangement of absorbers, a large dipole magnet, and fourteen planes of tracking and triggering chambers.

Observing the 'Primordial Matter' relics

The ultimate goal of ALICE is to track the remnants of the primordial matter, a task that requires the combination of four very challenging detection capabilities:

- *reconstruction* of all the tracks of tens of thousands of particles;
- *measurement* of the momentum of these particles from very low (100 MeV/c) to very high (≥ 100 GeV/c) values;
- *identification* of most particles through their specific interaction with different detectors;
- *observation* of the decay vertices, a fraction of a millimeter away from the collision, of the tell-tale heavy *charm* and *bottom quarks.*

Figure 36 shows a typical example of how the primordial matter might reveal itself in the ALICE tracker. The key to success is the combination of state-of-the-art tracking with specially developed low-mass silicon detectors and a very low-mass time-projection chamber optimized for this high multiplicity environment, followed by a suite of detectors specialized in identifying the particles. The different technologies are described in the following sections, beginning with the detectors closest to the point of (inter)action.

36. A simulated high-multiplicity event detected in the Alice tracking detectors.

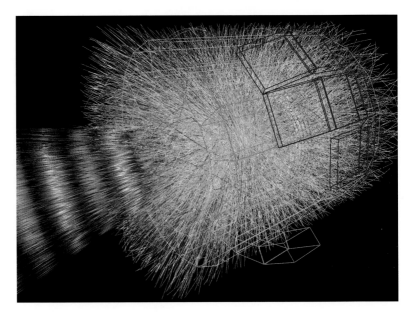

Tracking detectors

Tracking in the central barrel is divided into the Inner Tracking System (ITS), a six-layer, silicon vertex detector, and the Time-Projection Chamber (TPC). The ITS locates the primary vertex (where the collision occurs) and secondary vertices (where some of the unstable heavy particles decay after a flight distance of some hundreds of micrometers) with a precision of the order of a few tens of micrometers. Because of the high particle density, the innermost four layers need to be high-resolution devices, i.e. silicon *pixel detectors* and silicon *drift detectors*, which record x and y coordinates for each passing particle. The outer layers are equipped with double-sided silicon *micro-strip detectors*. The total area covered with silicon detectors reaches 7 m^2 and includes almost 13 million individual measurement channels.

The Silicon Pixel Detector (SPD)

The LHC experiments pioneered this novel detector technology. This is a 'checker-board' detector with tiny detection elements, typically 0.05 mm by 0.5 mm, resulting in a huge number (10 to 100 million) of detection channels. These detectors offer the best particle tracking capability of all presently existing methods. A host of problems had to be solved: each pixel element was individually connected to an equally tiny amplifier requiring the development of novel connection techniques ('bump bonding'); dedicated electronic microchips were designed to amplify and digitize the signal and to serialize the data; the detectors had to be cooled with the constraint of low mass; special low-mass support structures had to be designed.

The ALICE SPD is assembled from ten interlocking carbon-fiber structures, carrying two layers of pixels, cooling pipes and readout connections. In our almost obsessive efforts to minimize material, the pixel-detector wafers were manufactured with a thickness of only 0.2 mm; the electronic readout chips were mechanically thinned – after the bump bonding – from 0.3 mm to 0.15 mm; novel readout buses were developed, replacing copper (the technology standard) with aluminum for the readout tracks, again aiming to minimize the material of the detector. The result has been the creation of the world's lowest-total-mass pixel detector!

The ALICE SPD has another unique feature, not found in the other LHC pixel detectors: it provides a trigger on charged particles within less than 900 ns, in time for the next-level trigger decision. It will be used as an *interaction trigger*, allowing us to select very unusual multiplicity configurations. One half of the installed SPD can be seen in Figure 37, predominantly showing the conical fan-out of services to avoid occulting detectors near the beam pipe.

The Silicon Drift Detector (SDD)

The particle density behind the SPD is still so high that the subsequent tracker must also provide unambiguous two-dimensional space points with pixel-like space resolution. Apart from the pixel technology, the *silicon drift chamber* concept is the only one that provides this feature; it was selected

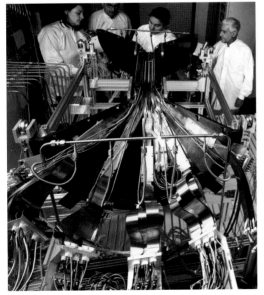

37. Partial view of one half of the SPD during installation. The service connections dwarf the actual pixel detector (photo: Antonio Saba).

by the ALICE collaboration as the best cost-performance option. This technology was never used on a large scale, although the original idea is more than 20 years old. Research and development had to address three main areas: the fabrication of suitable Si-drift detectors, the readout electronics and connectivity, and stability of operation. The relatively large size (88 by 73 mm²) of the 0.3 mm thick detector units contributed significantly to the production difficulties. A total of 260 such modules are installed in the completed detector. The electrical and cooling connectivity posed again one of the major technological hurdles; low-mass cables were *de rigeur*, implying the development of aluminum signal tracks on Kapton. This work, carried out in the Ukraine, was almost "torpedoed" by over-zealous customs officials, who decided that the readout buses were highly sensitive military material! As drift detectors are very sensitive to temperature variations, they will be thermo-stabilized to within a fraction of a degree and (in a belt and suspenders approach) monitored by numerous electronic structures distributed densely over the surface of the detectors.

Time Projection Chamber (TPC)
The need for efficient and robust three-dimensional tracking was the main motivation for selecting the time-projection chamber as the main tracking instrument. In spite of its drawbacks of slow recording speed and huge data volume, the TPC offers the advantage of providing highly redundant information, thus guaranteeing reliable performance with tens of thousands of charged particles within the geometrical *acceptance*.

A conceptual view of the TPC is shown in Figure 38. Charged particles leave ionizing tracks in the huge cylinder filled only with gas, which is separated into two halves by the central electrode. The electrons produced in the ionization of the gas slowly drift in a strong electric field of 100 kV towards the two ends of the TPC, where they are amplified and recorded in wire chambers (only one out of 72 of these chambers is shown on the left side of the figure). The TPC measures the many space points along each track; the arrival point in the chamber gives two coordinates, whereas the distance of the track from the endplate is inferred from drift time. A TPC is thus ideally suited to disentangle the dense web of particle tracks in heavy ion reactions.

This world's biggest TPC is the "workhorse" of ALICE; even by itself, the data it will provide will have a tremendous impact on our understanding of physics. Again, ALICE aimed for an exceptionally low-mass detector by mak-

38. Conceptual view of the TPC, showing the dimensions and components.

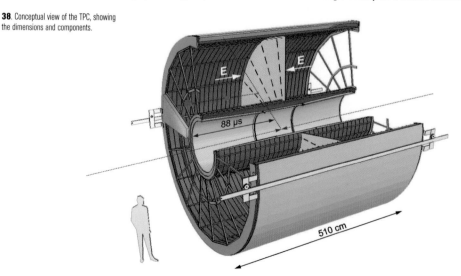

88 µs

510 cm

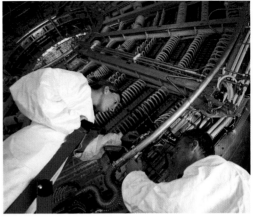

39. Inner tracking volume of the TPC. The layered structure provides a constant electric field of 400 V/cm between the central electrode and the readout chambers (photo: Michael Hoch).

40. Installation of the TPC readout electronics (photo: Antonio Saba).

41. View of the TPC during the installation; the inner tracking system is seen in its operating position.

ing the appropriate choice of materials and gas, combined with tight dimensional and environmental controls. The 5.6-m diameter and 5.4-m-long field cage is built from two carbon-fiber honeycomb-composite cylinders, materials normally used for space-applications. Dimensional tolerances are critical for the performance, such as the uniformity of the drift field; these were kept at the 10^{-4} level of precision.

The central drift electrode was built with a planarity and parallelism to the readout chambers of better than 0.2 mm, see the artistic view of the 90 m^3 volume of the inner chamber (Fig. 5) The second basic component to this ultralow-mass TPC is the chamber gas: conventional argon was rejected as unacceptably massive in favor of the much lighter neon. The final TPC is the lightest ever constructed, representing approximately the same amount of material crossed by the particles as the ITS.

We worked hard to develop innovative readout electronics: a preamplifier/signal shaper, operating at the fundamental thermal limit of noise, is followed by a specially developed readout chip, the ALICE Tpc Read Out (ALTRO) chip. It digitally processes the signals for optimized performance at high collision rates. The electronics is miniaturized to a level that allows it to be fully mounted on the end plates of the TPC, connected merely by 260 optical fibers to the data-acquisition system (Fig. 40). This readout is rapidly becoming the de-facto standard for gaseous detectors.

The performance advantage of Neon as the drift gas is not for free: environmental conditions (temperature and pressure) have to be controlled tightly, given the strong dependence of drift velocity on pressure and temperature. The needed 10^{-4} control on drift velocity requires a temperature stability of $\sim 0.1\,°C$ over the volume of 90 m^3. Needless to say that extreme thermal stabilization measures were employed to reach this goal. Figure 41 shows the TPC during the installation.

Particle identification detectors (PID)

Particle identification over a large part of the phase space and for many different particles is an important design feature of ALICE, and several PID systems are now operative:

- the Time-of-Flight (TOF) array measures the flight time of particles from the collision point out to the detector; together with the momentum this time determines particle mass;
- the High Momentum Particle Identification Detector (HMPID) uses *Cherenkov radiation* to measure particle velocities very close to the speed of light;
- the Transition Radiation Detector (TRD) will identify electrons above 1 GeV to study production rates of *heavy quarks* (charm and beauty mesons).

In addition to these dedicated detectors, the measurement of energy loss provided by the tracking detectors ITS and TPC is also used to indirectly identify particles; the former is well adapted to low momentum measurements (< 1 GeV/c), while the latter extends to very high momenta (> 60 GeV/c).

The Time-of-Flight System

The Time-of-Flight (TOF) System is designed to distinguish between pions, kaons and protons, with energies up to several GeV/c. The PID performance required a time resolution of the order of 50 picoseconds (1 ps = 10^{-12} s), in the presence of the extreme particle multiplicities. Comparable performance had been attained in small systems with scintillators, a solution that is neither practical nor cost efficient for the 150 m² of detector area and more than 150,000 detection channels. A quantum jump in technology was obviously required. Initially, several different methods were explored, but none was retained. Finally a breakthrough was reached with one venerable technique – the *resistive plate chamber* – through careful study of the detection process, insight and per-

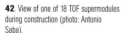

42. View of one of 18 TOF supermodules during construction (photo: Antonio Saba).

severance. The detector is made from 10 detection gaps of 0.25 mm each, delivering a staggering performance for time resolution of 50 ps, a factor 50 improvement compared to conventional RPCs. The high channel number required the development of novel microelectronics for the amplification and time measurement, with a precision of 25 ps. A new TOF standard was born and has already begun to revolutionize particle identification (Fig. 42).

The High Momentum Particle Identification Detector (HMPID)
The evolution of another method, to which many groups have contributed over the past 30 years, was pushed to new levels of performance: the *Cherenkov detector*. If, while traversing a medium (e.g., a gas or liquid), particles move at a speed that exceeds the speed of light, the particles excite the medium to emit *Cherenkov light*, named after the discoverer of this phenomenon. This light is emitted at a characteristic angle, determined by the velocity of the particle and the refractive index of the medium. By detecting the very faint light and its direction, the velocity can be determined, which allows deducing the mass, and hence the type, of the particle. This is the task of the high momentum particle identifier. The concept is as beautifully simple as it is fiendishly hard to turn it into a practical detector. The difficulty resides in detecting the very few (20 to 30) emitted photons through the use of detectors that can register the position of the photons with millimeter accuracy. The HMPID collaboration solved this problem by perfecting the production of large-area photocathodes, based on cesium iodide (CsI) films, evaporated onto gas-detector electrode surfaces, each about a square meter in size. A photon impinging on the CsI surface will produce a detectable electron with high probability (approx. 25 %). This development provided the basic ingredient for one of the world's largest imaging Cherenkov detectors ever built for an accelerator experiment, as seen in Figure 43.

43. The HMPID prior to installation in the experiment (photo: Antonio Saba).

The Transition Radiation Detector

Surrounding the TPC are 18 modules of a novel Transition Radiation Detector (TRD) that has the capability of distinguishing between ultra-relativistic particles (such as the very high-energy electrons) and the more conventional ones, e.g. pions. This detector has to perform with collisions that produce up to 20,000 particles; furthermore, ALICE must have a fast selection capability (under 6 μs) for electron and pion candidates; this is ground breaking technology for this type of detector.

The high multiplicity required a highly granular detector, and in this case the TRD is constructed of detection elements ('pads') of 6 cm^2 on average. The development phase addressed low-mass construction methods and the understanding of complex detector physics. The second area of development focused on improving the readout electronics, most of which had to be integrated directly on the back of each of the 540 detector layers. This detector comprises a staggering 1.2 million channels spread over 700 m^2, by far the largest TRD ever constructed and possibly one of the most complex LHC detector systems (Fig. 44).

Calorimeters

Photons, spanning the range from thermal radiation to hard quantum chromo-dynamic processes, as well as neutral mesons, are measured in the small single-arm, high-resolution and high-granularity PHOS electromagnetic calorimeter. PHOS, which literally will 'take the temperature' of the collision, is located far from the vertex (4.6 m) and is made of dense scintillating crystals (PbWO$_4$) in order to cope with the large particle density. While otherwise very similar in design to the CMS crystal calorimeter described earlier in this Chapter, it is cooled to −25 °C during operation to generate more light per incident energy, and therefore to improve the energy resolution. A set of multi-wire chambers in front of PHOS helps to separate charged particles from photons (CPV).

High-energy *partons* kicked out by hard collisions will 'plow' through the primordial matter, loosing energy along the way, before fragmenting into a spray of particles collectively called a *jet*; these modified jets therefore give information about the density and composition of the hot reaction zone. In order to enhance the capabilities for measuring jet properties, a second electromagnetic calorimeter (EMCal) has been installed in ALICE during a second phase in 2008. The EMCal is a lead-scintillator sampling calorimeter with longitudinal wavelength-shifting fibers; the signals are read out via avalanche photo diodes. Much larger than PHOS, but with lower granularity and energy resolution, it is optimized to measure jet production rates and jet characteristics in conjunction with the charged particle tracking in the other barrel detectors.

Forward and trigger detectors

A number of small and specialized detector systems are used for event selection or to measure global features of the reactions. Some of these detectors are shown in Figures 45 and 46 and their location in ALICE can be traced in Figure 35 with the help of the acronyms provided below:

- The collision time is measured with extreme precision ($< 2 \times 10^{-11}$ s) by two sets of 12 *Cherenkov counters* (fine mesh photomultipliers with fused quartz radiator) mounted tightly around the beam pipe (T0).
- Two arrays of segmented scintillator counters (V0) are used to select interactions and to reject beam related background events.
- An array of 60 large scintillators (ACORDE) on top of the L3 magnet will trigger on cosmic rays for calibration and alignment purposes, as well as for cosmic ray physics.
- The Forward Multiplicity Detector (FMD) provides information about the number and distribution of charged particles emerging from the reaction over an extended region, down to very small angles. These particles are counted in rings of silicon strip detectors located at three different positions along the beam pipe.
- The Photon Multiplicity Detector (PMD) measures the multiplicity and spatial distribution of photons in each single heavy ion collision. It consists of two planes of gas proportional counters with cellular honeycomb structure, preceded by two lead plates to convert the photons into electron pairs.
- Two sets of small, very compact Zero Degree Calorimeters (ZDC) are located far inside the LHC machine tunnel (> 100 m) and very close to the beam direction to record neutral particles emerging from heavy ion collisions in the forward direction.

45. One of the eight sectors of the V0A trigger hodoscope is shown here illuminated with UV light. The light of the scintillator pieces is piped out to the photomultipliers with wave-length shifting fibers that glow blue-green in the picture.

46. A head-on view of the Zero Degree Calorimeter (ZDC).

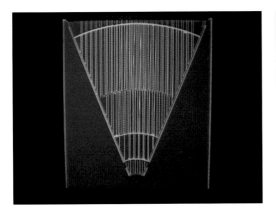

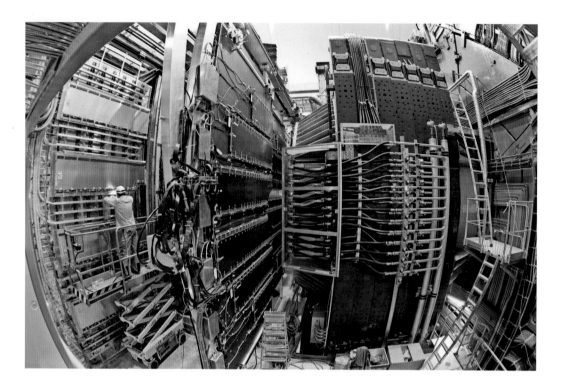

Muon Spectrometer

Located at one side behind the central solenoid and at the small angles between 2° and 9° relative to the beam direction is a muon spectrometer with its dipole magnet (4 MW power) generating a maximum field of 0.7 tesla. Several passive absorber systems (a hadron absorber close to the interaction point, a lead-steel-tungsten shield around the beam pipe and an iron wall) shield the spectrometer from most of the reaction products. The penetrating muons are measured in 10 planes of *cathode pad tracking chambers*, located between 5 m and 14 m from the interaction, with a precision of better than 100 μm. Again, the relatively low momentum of the muons of interest and the high particle density are the main challenges; therefore each chamber has two cathode planes, from which a double read-out provides two-dimensional space information. The chambers are made extremely thin and without metallic frames. The individual cathode pads range in size from 25 mm² close to the beam up to 5 cm² further away and cover 100 m² of active area with over 1 million active channels. Four trigger chambers are located at the end of the spectrometer, behind a 300-ton iron wall, to select and trigger on pairs of muons from the decay of heavy quark particles. The chambers are made using the *resistive plate* technology widely adopted by LHC experiments (Chaps. 5.2 and 5.3), and of modest granularity (20,000 channels covering 140 m²). Figure 47 shows a "fish eye" view of parts of the muon spectrometer in early 2008.

In order to select the few muons from the thousands of other particles produced in each collision, an absorber is placed in front of the muon spectrometer. As the name implies, it should very effectively absorb hadrons, while allowing muons to pass without much scattering (the usual task of such a hadron absorber), but also at the same time minimize the sideways flux of

47. Fish-eye view of the forward muon spectrometer; visible are (from right to left) the main solenoid (red) and the muon dipole magnets (blue), one of the tracking chamber planes and one of the triggering stations.

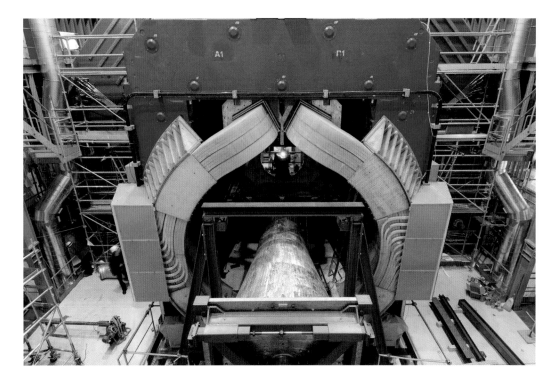

48. The conical hadron absorber slowly being pushed through the magnet of the muon spectrometer towards its final position.

shower debris (photons and neutrons in particular) which would harm the TPC performance. This seemingly innocuous object, seen during installation in Figure 48, is in reality the most complex and finely tuned particle absorber ever built: it weighs 40 tons, has a length exceeding 4 meters and protrudes into the TPC closely to the collision point. A composite structure was developed, consisting of an inner tungsten core surrounding the beam pipe, followed by a conical high-density carbon absorber in the acceptance of the spectrometer to minimize multiple scattering of muons, surrounded by a lead layer absorbing photons and an outer mantle of boron-loaded polyethylene to absorb neutrons. The whole assembly was literally 'cast in concrete' by pouring in mortar to fill out less critical spaces. From an engineering point of view, performance-cost optimization was a key issue, with close attention paid to sub-millimeter tolerances, fabrication and assembly procedures.

Fitting the pieces together: engineering and detector integration

Designing and constructing novel particle detectors was only part of the problem; building the mechanical supports and tools, fitting in the services and cables and, finally, assembling the pieces together was another critical challenge – and at first highly underestimated.

One example of sophisticated mechanical engineering in ALICE is the "space frame", a 9-m-diameter tubular stainless-steel structure that houses the combined 80 tons of the ITS, TPC, TRD, TOF and HMPID (Fig. 49). The collaboration insisted on a 'mass-less' structure maximizing active detector area and minimizing particle showers created in the support. During six months of optimization its weight was reduced by a factor of 2, while at the same time improved understanding of the detectors raised their weight by 50%! The resulting, complex, welded tubular structure deforms by up to 12 mm when

49. The space frame during installation. This stainless steel skeleton supports 80 tons of detectors, stressing its trusses close to the elastic limit. The mechanical integrity was load-tested on the surface prior to final installation.

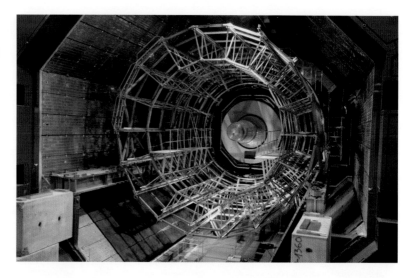

loaded. Laser and CCD camera angular monitors (36 in total) were specially developed and installed to provide a permanent monitor of the deflection and the feedback for adjustments.

Two features of ALICE, not found in the other LHC detectors, are at the origin of several unusual installation challenges:

- the asymmetric layout with the large magnetic muon spectrometer constrains the installation of and access to the central detectors: the 200 tons of 'central' detectors can only be installed from the opposite side;
- these central detectors placed inside the L3 magnet can only be supported from the mechanically rigid iron crowns at the ends of the magnet, separated by a distance of 15 meters.

These seemingly innocent issues, discussed below, kept several brilliant and creative engineers occupied for the better part of five years.

The asymmetry imposed by the muon spectrometer on the overall design was not accepted lightly. The alternative implied however displacing the 900-ton muon dipole magnet, the 300-ton muon filter, the muon detectors together with a large section of the delicate Be-vacuum chamber. The final verdict was rather clear: the complex, delicate one-sided installation represented the lesser evil.

First, the muon spectrometer was installed in its final, fixed position together with the hadron absorber. The installation of the inner tracking system and time projection chamber required a "ballet" to be minutely orchestrated, allowing the installation of the vacuum chamber, the independent pixel and ITS detectors and, finally, the TPC. Connecting the detectors to cables, gas and cooling lines required the team to place these detectors at various intermediate positions. This was not only complex and delicate, it was potentially dangerous: the movement of the detectors caused significant deformation of the supports (up to 5 mm in the vertical), while the vacuum chamber, attached to the detectors, was limited to excursions of less than 2.5 mm. This installation scenario could literally "make or break" the ALICE experiment; it was reviewed by many committees, dress-rehearsed on the surface with many of the final components, monitored with strain gauges, feeler gauges, cross-checked by survey teams, engineers and physicists, etc. It took the better part of nine months in 2007 to install these systems. Using a variety of tools from the inevitable duct tape to a dentist's drill for final dimensional adjustments, the operation was completed successfully and in time, as can be seen in Figure 50 which shows the experiment in early 2008, essentially ready to accept beams from the LHC.

Looking forward

New technology, skillful engineering, and critical design decisions have led to a state-of-the-art detector that will be up to the task of observing the primordial matter created in heavy ion collisions at the LHC. ALICE is the first truly universal "general purpose" heavy-ion experiment, which combines in a single detector most of the capabilities assigned in the past to several more specialized experiments. ALICE has incorporated the fruits of many years of research and development, dedicated specifically to meeting the numerous challenges posed by the physics of nuclear collisions at the LHC, and it is ready and well prepared, after more than 15 years of design and construction, to explore the "little bang" and enter ALICE's wonderland of physics at the LHC.

50. Front view of the L3 magnet, with its doors partially open. A bridge guides services, power cables, fluids and gases into the central detector. The silicon tracker is no longer visible and even the large TPC is mostly obscured. The stainless steel space-frame structure which supports most of the central detectors is partially filled with TOF and TRD modules.

Further reading
[1] K. Aamodt et al. (ALICE Collaboration) , "The ALICE experiment at the CERN LHC" *JINST* **3** (2008) S08002.

The Experiments

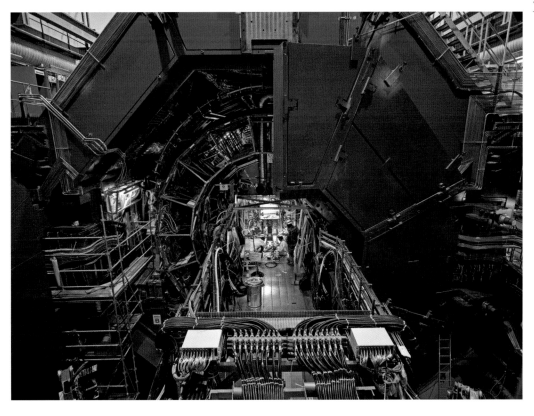

5.5

THE EXPERIMENTS

LHCb Experiments

Tatsuya Nakada

In contrast to the two "all-purpose" particle detectors for proton collisions at the LHC – ATLAS and CMS – the LHCb experiment was designed to answer a particular set of important open questions in particle physics. Particularly, the experiment will explore the mechanisms behind the preponderance of matter over antimatter in the universe that had already existed after the first milliseconds following the Big Bang.

The letter "b" in LHCb is a reference to beauty quarks (and their anti-particles), which are constituent of particles referred to as *B mesons*. Already 20 years ago, it was apparent that the high energy proton-proton collisions of the LHC would produce unprecedented amount of B mesons, and that should allow us to look indirectly for "physics beyond the Standard Model," as discussed earlier in the book (Chap. 2). For reasons that are beyond the scope of this discussion, the "b-quark sector" holds particular promise for entering into this domain of New Physics, through studying carefully phenomena such as *CP violation* and rare decays of a variety of B mesons.

The energies of the LHC collisions will produce a pair of b and anti-b quarks mostly emitted mostly in a direction very close to the beam axis; thus, instead of having a quasi-cylindircal geometry – such as ATLAS and CMS – which is optimized for particle detection perpendicular to the beam axis, the geometry of the LHCb resembles that of a reclined pyramid, with the apex located at the collision point. The LHCb experiment's 4,500-ton detector has been designed to efficiently detect B mesons produced by those b and anti-b quarks and to study the products of their decays. So instead of having the "onion-skin" structure of the other LHC particle detectors, the LHCb experiment stretches out 20 meters along the beam pipe, with its detectors spatially organized to efficiently measure the particles of interest.

In this chapter, we will present some brief history of the LHCb collaboration, followed by a description of the sub-detectors of the experiment. By taking this tour of the LHCb, the reader will see some specific characteristics of the detector optimized for *flavor physics*, so called because it mainly involves transitions between quark families, referred to as flavors.

Historical development of the LHCb experiment

The first achievements in the b-quark studies were largely achieved using electron-positron storage rings, in particular those operated at energies near 10 GeV, where pairs consisting of the B meson and its antiparticle were produced exclusively. The most recent machines called PEP-II at SLAC (Stanford, USA) and the KEKB at the Japanese KEK laboratories are even referred as "B factories" for their high running luminosities, which are capable of producing 10 B anti-B pairs per second without any other particles in the event.

From the early stages, the potential of the LHC project in abundantly producing b-quarks was clear. In fact, at the LHC energies, the production cross-section is estimated to be many orders of magnitude larger than that at the B factories. Success with B mesons achieved at the Collider Detector at Fermilab (CDF) experiment in 1992, also on a proton accelerator, increased the interest to design a dedicated b-physics experiment at LHC, and three Expressions of Interest (LoI's) were presented in March 1992. They adopted three different experimental approaches: COBEX was based on the colliding mode; GAJET based on the fixed-target mode using an internal gas jet target, where the detector would be built at one of the straight sections of LHC. The third proposal, LHB, also adopted the fixed-target mode but with an extracted beam obtained from the LHC beam hallow. Hadron identification and trigger capabilities were the two most important issues in the designs of these experiments. It should be noted that the SLAC and KEK B factories were approved only in 1993: Therefore, their physics goals were still focused on testing earlier premises concerning the mechanism of CP violation; since then, our understanding of these processes have evolved.

After the LoI presentations, three Letters of Intent were submitted in October 1993. Although the construction of the two B-factories had started by that time, Large Hadron Collider Committee (LHCC) recommended in January 1994 to foresee a dedicated b-physics detector at the LHC as one of the basic experiments since the large statistics available at LHC would still produce physics beyond the B-factory experiments. For this reason, LHCC asked the proponents of the three Letters of Intent to form a single collaboration in order to design a detector exploiting fully the potential of LHC based on the collider mode. By February 1998, a Technical Proposal of the LHCb experiment was submitted and accepted a few months later by the CERN management.

Since its approval, b physics has made a number of advances thanks to the B factories and Tevatron. Now the physics goal of LHCb has shifted from "testing of the Standard Model" to the "search for physics beyond the Standard Model." However, apart from the reduction of the number of tracking stations in order to reduce the material budget, no major change has been made to the detector design, and the LHCb experiment has successfully adjusted to this shift in scientific focus. The large production cross-section of b quarks at the collider mode remains the key feature. In addition, the trigger design evolved to rely more on software algorithms, giving us additional flexibility in selecting our scientific goals.

The LHCb Spectrometer

Requirements

The choice of the subsystems for the LHCb spectrometer was optimized for the physics goal. For example, it looks for a very rare process where a B meson decays into two muons with branching fractions expected by the Standard Model as small as 10^{-9}. A manifestation of "new physics" could increase this by some significant factor. In order to discover such rare decays, the detector must have the capability, for example, to measure momentum of the particles with extreme accurately and to identify muons among the decay products. This requires a very good tracking system and a muon system.

Due to the finite lifetime of B mesons, their decay points, so called secondary vertices, are away from the proton-proton collision point, usually referred as a primary vertex, where they are produced. For LHCb physics, the distance between the two vertices (in average about 7 mm) must be measured very accurately. For accurate reconstruction of vertices, trajectory of particles from those vertices must be well measured. This requirement calls for a *vertex detector* surrounding the proton-proton interaction point.

Let us also consider the requirements for a CP-violation study case. As we mentioned above, one of the main tasks of the LHCb will be the accurate measurement of CP violation, which must be studied in many different B meson decay channels. In some of these decay scenarios, the decay product contains a particle called a *kaon* (a member of the hadron family) or a proton. Thus, for identifying different decay channels, it will be essential to be able to identify the individual hadrons. For the same reason, we need to identify electrons, photons and muons. This requires a complex particle identification system.

Finally, as for the other experiments, we have strict requirements for the trigger. The fraction of events containing a pair of b and anti-b hadrons in the proton-proton interaction at the LHC energies is estimated to be an order of

51. (a) A photograph of the LHCb installation; and (b) view of the LHCb beam pipe as it passes through the central part of the detectors.

a

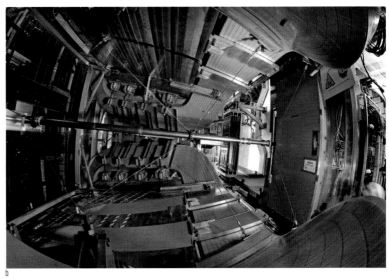

b

The Experiments

a few out of every 1000; our goal is to collect these efficiently out of the set of collisions occurring every 25 ns. One of the signatures of b-hadron events is the existence of particles whose transverse component of the momentum with respect to the beam axis, the *transverse momentum*, p_T, is greater than that of particles produced in the majority of proton-proton interactions. A fast p_T measurement can be carried out by the muon system, where the number of particles is small since most of the hadrons are absorbed by the hadron filter. Another fast p_T measurement can be done by the calorimeter system if the granularity is small enough for a given particle density. For this purpose, in addition to the electromagnetic calorimeter that measures the energy of electrons and photons, a hadron calorimeter is needed.

Based on these scientific goals, the LHCb spectrometer should have a vertex detector surrounding the interaction region, a tracking system with a magnet, a hadron identification system, a calorimeter system consists of a electromagnetic and hadron calorimeter, and a muon system.

Detector Layout
Although it is a proton-proton collider experiment, the layout of the LHCb detector shown in Figure 52 resembles a typical *fixed-target spectrometer* with its forward geometry. Here we will briefly mention the rather dense configuration of detection components that will be discussed in more detail in the next section. Starting from the proton-proton collision point and moving outwards we find:

- the vertex detector (VELO) made of a fine pitch silicon micro-strip detectors;
- the first Ring Imaging Cherenkov Counter (RICH-1) a particle detector based on the generation of Cherenkov radiation;
- the first tracking station made of silicon strip sensors (TT);
- a large-aperture normal conductive dipole magnet;
- three tracking stations (T1 to T3) each consists of a detector based of straw drift chambers for the most of the acceptance region (OT) and that based

52. A scale drawing of the LHCb experiment, showing the sequence of detectors laid out along the beam axis.

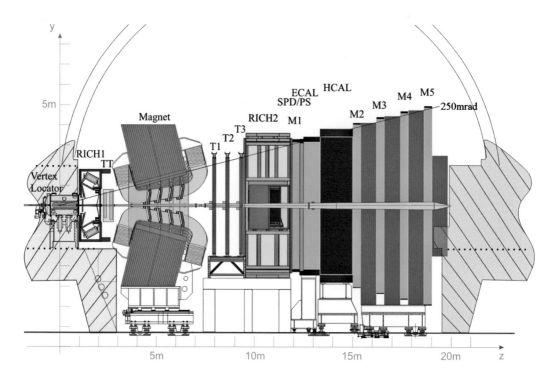

on silicon strip sensors for the region around the beam pipe (IT) where the track density is high;

- the second Ring Imaging Cherenkov Counter (RICH-2);
- the first tracking station for the muon (M1);
- a layer of scintillator pads (SPD) in front of a lead sheet and another set of scintillator pads behind (PS);
- the electromagnetic calorimeter (ECAL);
- the hadron calorimeter (HCAL);
- the four remaining tracking stations for the muon (M2 to M5) interleaved by iron absorbers.

As discussed previously, the choice of the detector geometry is based on the fact that both b hadrons and their antiparticles are predominantly produced in the forward (or backward) direction. The spectrometer can be built in an open geometry with an interaction region that is not surrounded by all the detector elements, as is the case for the ATLAS and CMS experiments. This allows a vertex detector system to be built with sensors that can be extracted away from the beam during the injection using so called *Roman Pot technique*. During beam injection, acceleration and collimation, the sensors are placed in a safe "rest" position away from the beam, while during the data taking, the sensors are moved much closer to the beam in order to achieve a good vertex resolution. In the forward geometry, most of the particle momentum is carried by the longitudinal component along the beam direction. The p_T threshold for the first level trigger can then be set to the value given by the trigger performance rather than by some detector limitation. The open geometry also allows easy installation, maintenance and possible upgrade of the experiment.

The LHCb detectors

For the remainder of the chapter, we will consider in more detail the individual detectors of the LHCb experiment.

The magnet

We begin our discussion of the detector elements with the magnet that supplies the field used to measure the momentum of charged particles. The LHCb magnet provides an integrated magnetic field path of 4 Tm for tracks going through the spectrometer. It has saddle-shaped coils in a window-frame yoke with sloping poles in order to match the required detector acceptance of 250 mrad vertically and 300 mrad horizontally (300 mrad = 17°). The yoke was constructed from relatively light slabs and coils were divided into 10 pieces. They were lowered down to the underground experimental area and assembled together at close to the final position. The total weight of the yoke is 1500 tons and of the two coils is 54 tons. The electric power of the magnet is 4.2 MW providing a peak field of 1.1 tesla with a current of 5.8 kA. The magnet can be seen in Figure 53.

53. The leading members of the LHCb magnet project, pictured in front of the installed magnet.

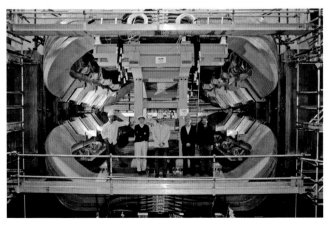

The vertex locator (VELO)

The vertex locator provides precise measurements of track coordinates

54. A schematic overview of the VELO detector.

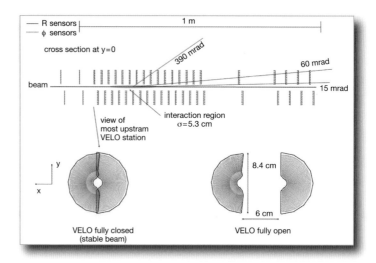

55. (a) A photograph of the assembly of the last module of the VELO system; and (b) a view of the same during installation into LHCb.

close to the interaction region, which are used to reconstruct the proton-proton interaction point, as well as the secondary vertices displaced from the primary vertex. As shown in Figures 54 and 55, the VELO consists of a series of micro-strip silicon sensors arranged along the beam direction. For accurate vertex reconstruction, the first measurement points of tracks must be as close as possible to the vertex. Therefore, the detector elements are positioned only seven millimeters away from the traversing proton beams.

Damage from the extremely energetic LHC proton beams is prevented with VELO's sensitive *Roman Pot* design feature. Its sensitive detector elements are mechanically retracted by 35 mm from the beam position while the beams are being injected and stabilized. Once the beams become stable, the silicon elements are moved back to the nominal position for data taking. This protects the VELO from possible damage during critical beam injection and dumping phases of the beam operation, and also permits the LHC operators to inject the beam without interference from the VELO detector.

The detectors are mounted in a vessel that maintains vacuum around the sensors and is separated from the machine vacuum by a thin aluminium sheet. This is done to protect the machine vacuum from possible contamination and to prevent the sensors picking up electric noises from the passing beams. The structure of the sheet is designed to minimize the material traversed by a charged particle before it crosses the sensors.

a

b

The silicon sensor has a half-moon shape and 0.3 mm thick. A small cutout in the centre of sensors allows the main LHC beam to pass through unimpeded. Charged particles produced by proton collisions traverse the silicon and generate electron-hole pairs; these electrons are sensed using application specific-electronics. The detectors and the read-out electronics are constructed from radiation resistant components, and a CO_2-based refrigeration system keeps the electronics cool. Signals from the detector are transported outside the vacuum system for analysis through 22,000 cables.

The Ring Imaging Chrenekov (RICH) detectors

Ring Imaging Cherenkov (RICH) detectors are used to identify a range of hadrons that result from the decay of B mesons, such as pions, kaons and protons. RICH detectors work by measuring emissions of *Cherenkov radiation*. This phenomenon occurs when a charged particle passes through a certain medium at a speed faster than that of light propagation in that medium. In fact, while the speed of light in vacuum is a constant, limiting speed in nature, when light enters a medium its speed is reduced by a factor known as the index of refraction. So an energetic particle can travel faster than light does in that medium, in which case a cone of light is produced, with a mechanism similar to the "bang" of supersonic planes as they exceed the speed of sound in the atmosphere. The opening angle of the cone depends on the particle's velocity, enabling the detector to determine its speed. Combining this information with the momentum measurement, the mass is calculated, which provides the identity.

In the both RICH detectors, the Cherenkov light is transported using a combination of spherical and flat mirrors to the photon detectors placed outside of the spectrometer acceptance (Fig. 56). Hybrid Photon Detectors (HPDs) shown in Figure 57 are used to detect the Cherenkov photons in the wavelength range 200-600 nm. The HPDs are surrounded by external iron shields and are placed in mumetal cylinders to permit operation in magnetic fields up to 50 mT.

The upstream detector, RICH1, covers the low momentum charged particle range 1-60 GeV/c using sililca aerogel and C_4F_{10} as the radiating medium. Silica aerogel is a colloidal form of solid quartz, but with an extremely low density and a high refractive index (1.01–1.10), which makes it a perfect radiator for the lowest-momentum particles. The downstream detector, RICH2, covers the high momentum range from 15 GeV/c up to and beyond 100 GeV/c and relies on CF_4 as the radiating medium. RICH1 covers the full LHCb acceptance from 25 mrad (1.4°) to 300 mrad (17°) in the horizontal direction, and up to 250 mrad (14°) in the vertical. RICH2 focuses on the high-momentum particles and covers the more limited angular acceptance of 15 mrad (0.9°) to 120 mrad (4.1°) in the horizontal, and 100 mrad (5.7°) in the vertical.

Particles produced in the collisions in LHCb will travel through the mirrors of RICH1 prior to reaching tracking system further downstream. To re-

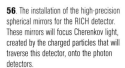

56. The installation of the high-precision spherical mirrors for the RICH detector. These mirrors will focus Cherenkov light, created by the charged particles that will traverse this detector, onto the photon detectors.

The Experiments

duce the amount of scattering, RICH1 uses special lightweight spherical mirrors constructed from a carbon-fiber reinforced polymer (CFRP), rather than glass. There are four of these mirrors, each made from two CFRP sheets molded into a spherical surface with a radius of 2700 mm and separated by a reinforcing matrix of CFRP cylinders. As RICH2 is located downstream of the tracking system and magnet, glass could be used for its spherical mirrors, which in this case are composed of hexagonal elements.

Tracking system

The LHCb tracking system consists of one station (labeled TT in Fig. 52) in front of the magnet, and three stations (T1 to T3) after the magnet. The principle task of the tracking system is to reconstruct trajectories of charged particles and measure their momenta. Tracking information of the charged particles is also useful for the reconstruction of Cherenkov rings in the RICH detectors.

The TT station, 150-cm wide and 130-cm high, surrounds the beam pipe and consists of four layers of silicon micro-strip sensors along the beam axis. The first and the last layer measure the track coordinate perpendicular to the magnetic field direction (needed for the momentum measurement), and the two other planes give a stereo view to obtain the other coordinate. All four layers are housed in a large light-tight detector volume that is both thermally and electrically insulated. In total, TT has some 270,000 readout electrodes and can measure the position of a particle to better than 0.05 mm.

The tracking layers T1 to T3 are made of the so-called straw-tube *drift chambers* (OT) covering a surface of about 30 m² per tracking station, except a small region around the beam pipe where silicon micro-strip detectors (IT) are employed because of the high particle density. Whenever a charged particle passes through the gas-filled straw tube, it ionizes the gas molecules, producing electrons. The position of the track is found by timing how long the electrons take to reach an anode wire situated in the center of each tube. Two staggered layers of straw detectors are glued together forming a super-layer in order to avoid the dead region between the two straws, and four super-layers along the beam direction with the same coordinate measurement scheme as

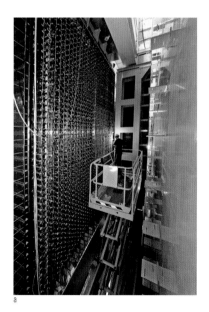

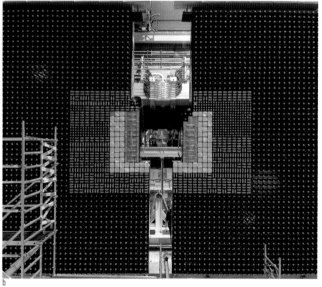

a

b

TT, constitute one station. The silicon micro-strip detectors consist of four detector boxes per station covering 0.35 m² of the surface surrounding the beam
pipe. Each box contains four layers of silicon micro-strip detectors, with the
same scheme of coordinate measurement as TT.

The calorimeter system

As mentioned above, the LHCb calorimeter system provides energy and position measurements of particles (electrons, photons and hadrons) produced
within its angular acceptance; this information is used for triggering and in offline event analysis. Furthermore, the electromagnetic calorimeter will be used
to reconstruct neutral pions which decay into two photons.

The LHCb calorimeter system consists of several layers: the scintillating-
pad detector (SP), the pre-shower detector (PS), the electromagnetic calorimeter (ECAL), and the hadron calorimeter (HCAL). The SP and PS systems of
scintillator pads are used to distinguish the electron from the photon, by
determining the presence or absence of electric charge, before the particle en-

ters the ECAL, where its energy measurement is performed. ECAL consists
of many alternating layers of thin
lead plates and scintillator plates. Its
signal is generated by particles colliding with its lead plates, which produces a shower of particles through
electromagnetic interactions. Those
particles then produce scintillating
lights when traverse the scintillator
plates (Fig. 59). The amount of light
is proportional to the energy of the
incident particle. Scintillation light is
collected by optical fibers and read
out by photomultiplier tubes. The
hadron calorimeter operates under

the same principle, but is larger and more dense, and uses iron plates as converters.

The muon system

The muon system is composed of five charged-particle tracking stations (M1-M5) of rectangular shape, interleaved by the hadron absorbers placed along the beam axis. The calorimeter system is used as the first layer of absorber, followed by the three iron block layers, each 80-cm thick, corresponding to a total of 20 *interaction lengths*. Another iron layer is placed after the last tracking station to shield the muon system against particles coming from the LHC tunnel. The minimum momentum of a muon required to cross the five stations is then approximately 6 GeV/c.

The muon system is used for two purposes: one is to provide a fast measurement of muon transverse momentum for the first level trigger; and the other is off-line muon identification for the physics analysis. For the first purpose, the granularity of the five stations, M1 to M5, is *projective*, meaning that all their transverse dimensions scale with the distance from the interaction point. This speeds up the reconstruction and momentum determination of muon tracks. Multi-wire *proportional chambers* are used to detect passing muons; each station contains chambers filled with a combination of three gases (carbon dioxide, argon, and tetrafluoromethane), and the passing muons ionize this mixture, producing electrons that can be detected by wire electrodes. The full system consists of 1380 chambers, covering a total area of 435 m^2 (about the size of a basketball court). The inner region around the beam pipe of station M1, where the expected particle rate exceeds safety limits for radiation damage of MWPC, is equipped with triple-gas electron-multiplier detectors.

The electronics is based on custom radiation-hard chips developed specially for the muon system. The detectors provide space-point measurements of the tracks, by supplying data in a binary (yes/no) form. A total of 126,000 front-end readout channels are used.

The LHCb trigger

The LHCb has adopted a two-level trigger system. The first level of the trigger reduces the LHC bunch collision rate of 40 MHz to an event rate of about 1 MHz, based on the p_T information as described earlier in this Chapter. The system was built with custom-made electronics, but based on commercial components. For those events resulting in a positive trigger decision, signals from all the detector subsystems are readout and transferred to the memory of one of the computers in the so-called event-filter farm, where 1000 to 2000 commercially available PCs are housed. The final event selection is done with software running on those PCs, and data for the selected events are written on a permanent storage at a rate of 2 KHz, to be used subsequently for off-line event reconstruction. The event rate reduction in the first level is rather modest, and the efficiency to maintain events with b-hadron activity is rather high. Since the second level is purely done by the software, the selection algorithms can be easily adapted to evolving physics requirement.

Looking forward to a few surprises from Nature

The LHCb experiment is the result of almost 20 years of work of an international team; this remarkable instrument – uniquely qualified to cast light on some of the most mysterious physical processes as they occurred in the first instants of the existence of our universe – is now installed and ready to take data.

60. An inside view of the LHCb muon detection system, showing two stations on either side and a green iron absorber block in the back.

Further reading
LHCb Collaboration Members,
"The LHCb Detector at the LHC"
JINST **3** (2008) S08005.

5.6

THE EXPERIMENTS

LHC Data Analysis and the Grid

John Harvey, Pere Mato,
and Les Robertson

When the LHC is operating it will produce an enormous amount of data. If all were recorded, it would amount to a data rate of 1 petabyte (1 PB = 10^{15} bytes – equivalent to 200,000 DVDs) every second. Special electronics embedded in the detectors and dedicated online computing systems will reduce this by six orders of magnitude to a more manageable 1 gigabyte (1 GB = 10^9 bytes) per second, which will then be recorded on magnetic storage for later processing. Even with the online reduction, a huge quantity of new data, growing at a rate of nearly 15 PB each year, has to be carefully managed and made readily accessible to the thousands of scientists around the world who are engaged in LHC physics analysis.

The algorithms required for simulation and processing of the data are very complex; these were developed by specialists working in the different specialty areas of the physics being studied, as well as by experts in the materials and technologies of the different sub-detectors. But the factors that differentiate LHC computing from previous high-energy physics experiments, and from most other scientific experiments are

- the enormous number of physicists and engineers participating actively in data analysis – the CMS experiment alone has more than three thousand members – a large fraction of which are involved in algorithm and program development;
- the widely distributed computing environment; about 100,000 processors installed in 140 computer centers in 35 countries are integrated into the LHC computing grid;
- the huge quantity of data that has to be distributed across the grid and shared by all of the members of each experiment.

The design of all aspects of the computing system – frameworks, programs, data models, computing services – began more than ten years ago, and all of these will continue to evolve over the many years of the operational lifetime of the LHC accelerator. Computing technologies have changed and will continue to change throughout this period, requiring close attention to flexibility of design and implementation. Effective collaboration between so many independently minded researchers is itself a major challenge.

This chapter describes some of the technologies that are currently being used, as well as some of the management and organizational approaches that have been adopted. Although the feasibility of this approach is established, many challenges remain to be confronted when the data begins to flow.

The data-processing model

To achieve the physics goals, sophisticated algorithms are applied successively to the raw data collected by each experiment in order to extract physical quantities and observables of interest that can be compared to theoretical predictions. Figure 61 shows a high-level view of the data flow and the principal processing stages involved in this process. As can be seen in the Figure, there are stages of event selection and reconstruction, before the data are exploited for physics analysis. Event simulation is also an essential part of any physics experiment, since it provides a better understanding of the experimental conditions and the performance of the detector, both in the design and optimization phase, as well as during real taking of data. The stream of simulated events is produced by simulating the passage of particles through the detector material to produce "hits" in the detectors which are then reconstructed and analyzed in exactly the same way as for real data. In this Chapter, we present an overview of the data processing model.

Our discussion begins at the four experimental interaction regions, where bunches of protons cross 40 million times per second. Each bunch contains 10^{11} protons and, in normal operation, the nominal rate of proton-proton collisions seen by each of the ATLAS and CMS detectors will be about one billion per second (10^9 Hz). If all the data generated by these collisions were recorded, this would amount to a data recording rate of 1 petabyte per second. In fact only a tiny fraction of these collisions involve physics processes that are of interest for further analysis. For example, a 100 GeV/c^2 Higgs will be produced at 0.1 Hz and a heavier one of 600 GeV/c^2 will be produced at 10^{-2} Hz. In all, the selection needed is one event out of 10^{10}-10^{11} events. This very strong selection is realized by a number of successive steps: the first steps are performed in real-time by the trigger and data acquisition systems, and the later ones are performed "off-line" by sophisticated data processing programs.

In computing terms the online selection process must reduce the initial data rate such that the storage and further processing of data offline are both manageable and affordable. In practice, online data are reduced by six orders of magnitude and are passed to the offline data handling system at an average

61. LHC Data Analysis – High level diagram of the data flow and the major processing stages.

rate of 320 MB per second (1.25 GB per second in the case of the ALICE experiment when running with beams of lead ions). An initial offline processing stage, *event reconstruction*, transforms the raw data into a physics view of the event, finding particle tracks and computing physical quantities such as position, time, momentum, and energy. Depending on the detection technique, the transformation from the digital output of the electronics to physical quantities can be as simple as applying a calibration, or it may involve sophisticated signal processing followed by complex *clustering* and *pattern recognition* algorithms. The raw data are also used to obtain a better understanding of the alignment and calibration of the detector elements; the event reconstruction is typically repeated several times to obtain improvements in the measurements of the tracks and energy deposits. The final steps in event reconstruction involve identifying particles (electrons, muons, jets and missing energy) and classifying the event according to its physics characteristics; in a way, this stage transforms the information from the level "there was a hit in this point of the detector" into "there is a likely electron with a certain energy pointing in this angular direction." The output from this stage, the *event summary data* (ESD), forms the basis for further stages of analysis.

Physics analysis, shown in green in Figure 61, is an iterative process, involving many passes through large numbers of events. A large fraction of the collaboration – several hundred physicists – takes an active part in the analysis, and so the data must be carefully organized to avoid access contention. Two factors help in this: firstly, each event is independent, and so there is ample scope for performing the processing in parallel; secondly the data are read-only with updates applied only to new versions of the datasets, and so replicas are easy to manage. File catalogues are maintained to allow analysis tasks to pre-select the events of interest. Access to the ESD is usually organized as a production activity with large numbers of parallel batch jobs trawling through major subsets of the data, generating extracts for specific classes of events according to the current analysis interests. These extracts are small enough to be replicated many times and serve as an input for the analysis activities of the large majority of the physicists, either interactively or in batch mode. At this level, end-user analysis, i.e., the way in which the data will be accessed, is hard to predict, as it will depend on the physics that emerges and the novel ideas that emerge for analyzing the data.

Simulation of the performance of the detector, shown in pink in Figure 61, is also extremely important for the optimization of the design and for understanding inefficiencies and resolutions of the real detector under data-taking conditions. *Monte Carlo techniques* (so called because they are based on random-number generation, like the roulette wheel at Monte Carlo) are used to generate physics events following a given theoretical model, and to simulate the electromagnetic and nuclear interactions of the stable particles composing the event in the active and passive materials of the detector. The response of the readout electronics is also simulated in order to produce raw data in a form that corresponds to the real data generated by the detector during data recording. Simulation software requires accurate descriptions of the geometry and material composition of the detectors and makes heavy use of common software toolkits for simulation of primary physics interactions, written by theoretical physicists according to the most recent models (e.g. Pythia and Herwig), and for detector simulation written by experts of particle interaction with matter (such as Geant4 and Fluka). Simulation programs are very demanding in memory as well as in CPU resources. However from a computing service standpoint, simulation is relatively easy to manage as it can be run as a background production activity and, being computationally intensive, the data rates are relatively low.

Software design and development

Major challenges

Managing the scale and complexity of the data from the LHC experiments provides the biggest challenge for the software and computing systems. Figure 62 compares the event size and trigger rates for a number of large HEP experiments that have been active over the last two decades. The LHC event size is an order of magnitude larger than in previous experiments. This results from the finer *granularity* of the detectors together with the complexity of the event itself in terms of the number of particles produced in collisions at the LHC energy scale. The trigger rate is also significantly higher, as already explained, and so the total data volume, which is the product of the event size and the trigger rate, is orders of magnitude greater than in previous experiments.

The complexity of the data processing and the resulting computational requirements is also proportional to the number of tracks in each collision. Sophisticated algorithms, such as pattern recognition, are needed to provide the correct identification and measurement of all particles with very high efficiency. The processing time for this kind of problem scales more than just linearly with the number of real particles.

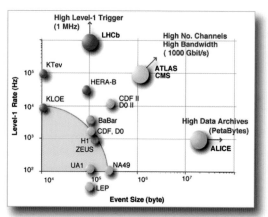

62. Comparison of the trigger rates and event sizes for recent HEP experiments and the current LHC experiments.

Another aspect that differentiates these new experiments from previous ones is the number of collaborators that contribute to the development of the software system. The programs for processing the event data are very specific to each experiment and require developers with a deep understanding of the detecting apparatus, their associated electronics and the physics. Several hundred people from the LHC collaborations have developed a total of *several million of lines of code* for each experiment. These software developers are typically physicists and detector specialists from different institutions and laboratories distributed around the world who work only part-time on software development and have a wide range of software engineering skills. Previous experiments, such as those at the LEP accelerator at CERN, had much smaller software systems produced by a few tens of developers working in much more centralized teams. The challenge has been to integrate the contributions of this diverse community of developers into a complete and coherent running system. New software development methods and technologies have helped enormously in meeting this challenge.

Software design and implementation

Software environments and technologies evolve over time and therefore software design must take account of the lifetime of the LHC, which will exceed 10 years. The experiment software itself must also be able to evolve smoothly with time as new requirements emerge. Accommodating change and long-term maintainability implies paying special attention to software qualities such as modularity and reusability, as well as the provision of documentation and training for the large communities of developers. At any given time the exper-

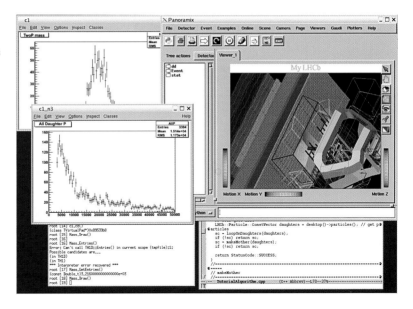

63. Typical interactive environment for a physicist in one of the LHC experiments in which he/she develops new data analysis algorithms and shows the results in terms of statistical distributions or the event display.

iments must provide a fully functional set of software with implementations based on products that are the current best choice.

The main programming language selected for the development of physics software applications in all four LHC experiments is C++, with some legacy algorithms written in FORTRAN, and Python used in cases where a scripting language is more appropriate. The *object-oriented* paradigm is ubiquitous. LHC software must be able to run seamlessly in a highly distributed environment and on a variety of platforms (hardware/operating system/compiler combinations), in several flavors and versions. Platform dependencies are typically confined to low-level system utilities. The set of supported platforms is reviewed periodically to take account of market trends and usage by the wider community.

The software development strategy follows an *architecture-centric approach* as a way of creating a resilient software framework that can withstand changes in requirements and technology over the expected lifetime of the experiment. The architecture consists of components with well-specified functionality, and interfaces that describe the way these components interact. Components can be reused in different configurations in order to support the full range of event data-processing applications: trigger, reconstruction, simulation and analysis. These configurations include the production environments used for the prompt processing of data as they are collected (high-level trigger and event reconstruction are examples), as well as the distributed computing environments utilized by physicists performing their individual analyses. The approach has proven to be very successful in hiding details of the complexity of the underlying technologies and in facilitating understanding and communication between developers.

Figure 64 shows a schematic view of the GAUDI architecture, which is used by LHCb and ATLAS and which is similar in many respects to the frameworks used by CMS (EDM) and ALICE (AliROOT). This architecture considers the algorithmic part of any data processing as a set of objects that are distinct from the objects holding the data of the event. This decoupling between the objects describing the data and those implementing the algorithms allows programmers to concentrate separately on each. It also provides for

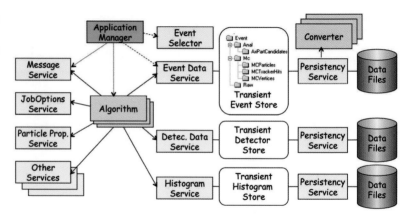

64. The Gaudi Architecture – An example of a software framework used by LHC experiments to provide the flexibility needed as the requirements evolve and to ensure independence of the underlying computing technologies.

longer stability of the data objects (the *event model*) as algorithms tend to evolve much more rapidly, while the enormous volume of data dictates a more conservative approach for the event model. A sequence of algorithms performs the actual data transformations from the detector raw data to sophisticated objects representing real particles.

These experiment frameworks are built on top of a number of more general purpose packages specialized in various software domains. There has been a long tradition of developing libraries oriented towards the more general needs of the world-wide HEP community as they foster the adoption of common solutions between the different experiments and help to harmonize the exploitation of the computing infrastructure worldwide. They typically provide the functionality needed to handle and analyze large amounts of data and include data persistency mechanisms, various database back-ends for accessing time varying detector data such as calibration or alignment constants, mathematical and statistical libraries, and 3D graphics libraries for supporting event visualization. They include many well-supported *open source* packages. While the use of external packages has significantly reduced the overall development effort for each experiment, there is also a down side – the increased complexity of configuration management and system integration, as well as the creation of dependencies on the communities developing these packages.

The software-development process
Quality requirements vary widely between the different software components, depending on the criticality of the applications in which they will be used. For example, algorithms developed for event selection in the online event filter farm must be very reliable, robust and error-free as there is no possibility to repeat the process once the event has been discarded. Sophisticated real-time monitoring is put in place to detect any malfunctioning of the trigger systems. On the other hand, applications that run on data stored offline can be rerun once imperfections are found and corrected, or after improved algorithms become available.

In addition to the standard data processing steps organised by computing specialists within the experiments, scientists from the LHC collaborations develop their own algorithms and programs to perform their own individual research.

Integrating the contributions from the large number of developers to form a complete working application, such as event reconstruction, has been a huge challenge. In the software industry strict software engineering practices are usually employed to manage the activities of large software production units. This approach is not easy to follow in the context of a scientific community where the majority of the developers are physicists and detector specialists who find writing code within the constraints of a formal software engineering environment to be irksome. Every experiment has therefore put in place a rather light software process that permits developers to work more or less independently and, at the same time, facilitates the integration of their contributions into the large programs. The bigger collaborations typically have more than 250 different developers committing code changes every week. The software is organized in packages with sufficient granularity such that very few developers work on a given package at the same time. The total number of packages has grown significantly with time and today each experiment has more than a thousand, each containing the implementation of a number of related classes. Once a certain level of functionality is reached, the package is tagged by the author and can be integrated into a given release of the software. A number of tools are used to manage the code repository, the collection of tags, the dependencies between packages, and the process used to build the release.

During the development phase it has been a major challenge to provide a stable environment for developers to write their algorithms as well as a comprehensive set of simulated data on which to exercise them. Each experiment has developed an automated system for integration and validation of new software builds that runs at least once per day. This is essential during data recording in the case that bugs are discovered and need to be fixed in a very short time. On the other hand, experience has shown that the time to converge to a validated release for the whole collaboration can take a very long time, illustrating the complexity of both the software development activity and the community of independent developers that need to collaborate. Before the software is released for use by the collaboration, it undergoes a series of regression tests and validation tasks are run to ensure that it is working correctly. In addition, the performance of the code is measured in terms of its usage of memory, CPU time and data storage, in order to map the applications to the resources available.

Base technologies and capacity requirements

Experimental high-energy-physics programs perform well on the general purpose PCs designed for office and home use. A simple distributed architecture was developed around 1990, taking advantage of the parallel nature of physics analysis, allowing the computing services to migrate from specialized scientific computers to inexpensive clusters that are today built from simple PC components. High-energy physics has therefore been able to benefit for almost twenty years from the mass-market-driven growth in the performance and capacity of processors, memories, disks, and local area networking equipment.

For the four experiments, the master data will grow at around 15 PB per year, but with intermediate versions and replicas the volume of disk storage that must be managed will grow at about three times this rate, all of which will require the computational capacity of about 100,000 processors in the first full year of operation. The volume of data and the need to share it across very large collaborations are the key computing issue for LHC data analysis (Fig. 65).

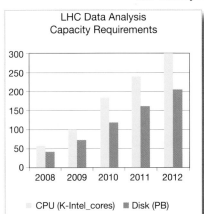

The LHC Computing Grid – a worldwide virtual computing Center

When work began in 1999 on the design of the computing system for LHC data analysis, it rapidly became clear that the overall capacity required for the initial four experiments was far beyond the funding that would be available at CERN. On the other hand most of the laboratories and universities that were collaborating in the experiments had access to national or regional computing facilities, and so the obvious question was: Could these facilities in some way be integrated with CERN to provide a single LHC computing service? The easy parallelism inherent in the analysis, together with the rapid evolution of wide-area networking – increasing capacity and bandwidth coupled with falling costs – made it look possible and this was confirmed by a feasibility study that developed the system architecture shown in Figure 66.

- CERN, as the "Tier-0", performs initial processing of the data and maintains master copies of the raw and other key datasets, pushing the data out rapidly to ....
- .... eleven large data-intensive centers – with major investments in mass-storage services, round the clock operation, excellent network connectivity. These are called the "Tier-1" centers, and provide for long-term data preservation, hold synchronized copies of the master catalogues, are used for the data-intensive analysis tasks, and act as data servers for smaller centers.
- The end-user analysis tasks, the heart of the physics discovery process, are delegated to about 120 "Tier-2" centers, located close to the end user, i.e., in large universities or research centers. While they do not have to make the same level of commitment in terms of data and storage management they must adapt their configurations to support the evolving demands of their client physics groups. The expectation is that diversity among the Tier-2s will stimulate novel approaches to analysis.

Tying all of these centers together to provide a coherent service was a major challenge. Rather than develop a special solution for LHC, it was decided to implement the distributed system as a *computational grid*, based on the ideas of two scientists working in the United States, Ian Foster and Carl Kesselman. They had developed software that allowed computing centers to inter-connect in a very general way, integrating their resources to offer a *virtual computing service*. The physicist sees a single service, enabling her to concen-

66. The LHC Computing Service Hierarchy, which evolved from the work of the Monarc project in 1999.

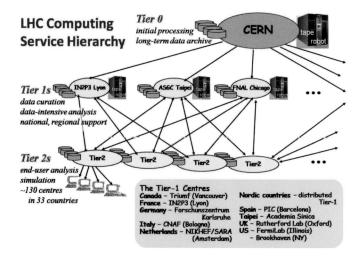

trate on her analysis without being troubled by the details of where the data is located, where the computational capacity is available, how to authenticate and how to obtain resource allocations from more than a hundred independently managed computer centers.

Grid technology

A site participates in a computing grid by running a set of services that enable other sites to see the resources that are available (the *information service*), submit work (the *compute element*) and access data (the *storage element*). A *security framework* provides the basis for trust between the sites connected to the grid, defining the rules for authentication of users and grid components using digital certificates. Users of the grid are grouped into *virtual organizations* (VO), typically one per experiment. An individual user is registered with the VO, and is then able to use resources and services at any site that supports that VO. Some of the resources in the grid may be owned exclusively by one VO, while others may be shared between several VOs. There is no central control over the resources and the configuration – each site advertises its resources and their VO affinity to grid users through the distributed information service.

Figure 67 shows the set of services provided by the *gLite* package that is used by sites that participate in the *Enabling Grids for E-sciencE* (EGEE) grid, the majority of the sites serving LHC. The other major grid used by LHC sites is the *Open Science Grid* (OSG), which uses a very similar toolkit. EGEE receives support from the European Commission and OSG is partly funded by the Department of Energy and the National Science Foundation in the US. In addition to providing a grid software package these *infrastructure grids* also coordinate the operation of their grids, providing services such as quality of service monitoring, problem determination, resource accounting and user support. High-energy-physics institutes were heavily involved in the predecessors of these grids, and remain major users, but the grids also serve many other sciences.

LHC uses an application-level grid built on top of a small number of infrastructure grids (EGEE, OSG, Nordugrid in the Nordic countries, Westgrid in Canada). LHC defines a set of *baseline services* that must be supported at all participating sites. Some of these are simply services provided by the under-

The Experiments

67. The Grid Middleware Services provided by the gLite Toolkit.

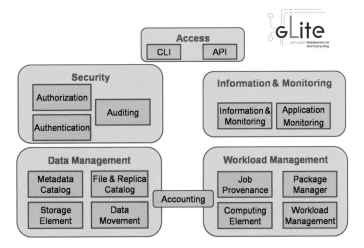

68. A Sun StorageTek L8500 Enterprise Magnetic Tape Library, as used at CERN and several of the Tier-1s, that can store tens of petabytes of data. Magnetic tapes are located all around the walls, and a small robot, visible at the bottom center, extracts and loads the tapes.

69. A view of the CERN computer center during the installation of the servers.

lying infrastructure grids, others have been developed specifically for LHC, some use proprietary software packages, and others are implemented in different ways at different sites but adhere to agreed standards. The grid concept provides a great deal of flexibility to the application, which can pick and choose the services that it requires. Flexibility of course also has a down side in terms of complexity and maintenance cost.

Data and storage management

The biggest challenge in constructing and operating a grid for LHC is managing the vast quantity of data – growing at about 45 PB per year – which has to be moved reliably around the grid, whether as part of scheduled production processes or driven dynamically by the needs of the analysis activities. CERN alone must sustain a long-term data export rate of some two GB per second, while the aggregate data rate at a Tier-1 may exceed one GB per second.

Distributed data management also involves managing the storage space at each of the sites. Several different storage management systems are used, depending on the specific requirements of each site (capacity, performance, functionality), but all of them support a standard set of functions for manipulating storage: specifying storage classes, allocating data spaces, naming files, initiating archive and recall, etc. The *Storage Resource Manager* functions are accessible from the grid, enabling applications and administrators to manipulate storage remotely using the same interfaces at all grid sites.

The most challenging part, however, is keeping track of the tens of millions of files generated by the applications, maintaining consistent distributed catalogues that define their location, the location of replicas and the metadata describing their status and physics content. Each of the experiments has invested heavily in software systems to manage their data and the storage available to them at the grid sites, and make the data available transparently to their users with the required performance. Figure 70 shows the architecture of the ATLAS experiment's distributed data management system. This is responsible for a wide range of functions including registering new files, creating replicas, managing metadata updates, ensuring consistency of catalogues, keeping track of available storage resources and migrating data between sites.

70. Architectural diagram of the ATLAS experiment's Distributed Data Management (DDM) system.

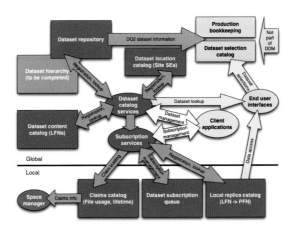

Network switching equipment installed at CERN to manage the thousands of fibres terminating in the computer centre

International networking

Reliable high-bandwidth international networking is essential in order to achieve the performance and reliability goals for LHC data transfer. For communication between CERN and the Tier-1s, and for inter-Tier-1 traffic, an optical private network (OPN) has been established. This provides point-to-point connections between CERN and each of the sites, implemented as a wavelength carried across an optical fiber and capable of transferring data at ten GB per second. High-speed network routers at the sites enable data to be switched between these and other inter-site links, providing redundant data paths that can be used in the case of individual link failure. An important benefit of the private nature of the OPN is that the bandwidth between sites is guaranteed, ensuring that the required end-to-end data rates can always be achieved.

High-energy physics has long been a major user of long range networks and has played a leading role in the development of the international research networking infrastructure which, in the large majority of cases, provides sufficient bandwidth for communication between Tier-1s and Tier-2s. CERN's data needs make it the largest single network user in Switzerland for research and education. As the optical-fiber infrastructure was developed over the past decade a great deal of fiber has been routed to the CERN site. This has enabled an exchange point for IP traffic and, more recently, an optical exchange point for circuit switching to be established. This gives CERN a great deal of flexibility to extend its network capability in a cost-effective manner as the bandwidth requirements grow and as new centers join the LHC grid.

The user interface

To complete the task of masking the complexities of the grid from the end user each experiment has designed its data management, resource scheduling and job preparation systems to deal with the distributed environment.

For example, the Ganga data-analysis framework, used by two of the experiments, enables the user to work in the same way whether running analysis across limited data samples on a local system or in parallel mode across very large datasets on the grid. The user specifies the algorithm and defines the data to be analyzed (e.g. as a query on a catalogue for a given set of data).

Ganga uses the experiment's data management system to establish the files that are required and their locations. Using this information and various application specific rules, it then splits the task into a number of independent jobs. When the jobs are completed, Ganga merges the outputs and returns the results to the user. In this way the user can move easily from testing an algorithm on a small dataset to applying it to the full dataset and does not need detailed knowledge of the configuration and status of the grid resources.

The distributed computing model

As we have seen, the computing service for LHC data analysis is implemented as a geographically distributed computational grid. Each of the experiments has developed a distributed *computing model* that maps the data and processes to the architecture of the grid in such a way as to facilitate the efficient exploitation of the available resources. The computing model allows the collaboration to locate data where they can be most efficiently processed and to manage the utilization of the resources available to the experiment at each grid site according to the experiment's priorities. This represents a departure from the *centralized model* used by previous experiments, where the major part of the resources was located at the accelerator laboratory located at CERN.

Complying with the computing model, the experiments have established initial policies for locating bulk experiment-wide data at sites according to static factors such as the classification of the event. This simplifies the scheduling of the workload, as jobs need only be steered to one of the locations at which a copy of the data has been stored. It is expected that the models will evolve to include a more dynamic movement of data in response to workload. The data management systems are implemented as a set of loosely coupled tools for describing datasets and their physical location (catalogues), for managing data transfers and for providing access to collections of events stored in files. The packaging of events into files is done in such a way that the average file size is kept relatively large (on the order of or exceeding one GB) in order to avoid practical scaling issues that arise with storage systems. Small files that are generated by individual jobs are merged into files of adequate size that can be tracked by the data management system.

Data processing is organized in terms of workflows that define the interactions with all necessary systems and services. Production software tools have been developed that manage the various phases involved in the execution of these workflows. Typically the user decides which application is to be run (reconstruction, analysis etc.) and the required configuration such as which input dataset to use. The experiment's data catalogues are then queried to find the location of data to be processed; portions of the dataset are assigned to different jobs and decisions are made as to where the jobs will be dispatched. Once jobs are completed, output datasets are transferred to the required destination site – this may be local or the output may be passed to the data transfer system for storage elsewhere on the grid.

One critical role of data management is to provide rapid feedback on the quality of data and to assure optimal detector operation. For this reason, the prompt reconstruction and calibration workflows typically run as soon as data are provided by the high-level trigger. The process involves the repacking of event data into different streams (*primary datasets*) according to their trigger signatures (i.e., the class of event to which it belongs as defined by the main particles present in the event). This procedure facilitates the access to data during subsequent processing phases. Calibration also uses dedicated streams of data taken with various special trigger sources. These and other latency critical workflows, such as express stream analyses, run in their entirety at CERN. Typically one copy of the raw and reconstructed data is stored at CERN and another copy is distributed amongst the Tier-1 centers.

The ultimate detector accuracy will be achieved by detailed studies and precise calibration procedures that will require processing a large number of events. These studies will use the results of the prompt reconstruction to produce calibration data for subsequent reconstruction passes, which can also take advantage of improvements made to the reconstruction algorithms. This workflow may be executed wherever raw data are located, i.e., either at the Tier-1 centers or at CERN.

Analysis is typically organized according to the analysis topic and uses datasets (skims) containing events with specific characteristics related to the final state sought by the physicist. The skims are representative of all the samples required to carry out an analysis and are used for designing and optimizing analysis procedures and code. They are stored close to the physicists at the Tier-2 sites and in local analysis facilities in universities and laboratories. Analyses requiring complete samples with high statistics can be run over all the desired primary datasets at the Tier-1 sites at which they are stored.

Simulation is a computationally intensive activity with reduced data bandwidth requirements, and can be executed at centers without mass storage facilities. Large productions of simulated events (Monte-Carlo simulations) are therefore run at Tier-2 centers and datasets of simulated events are transferred to the associated Tier-1 for data archival and analysis.

The computing model must evolve to take account of changes in the data processing requirements with time. For example, initially event data will need to include a substantial amount of extra information needed to understand detector performance and to tune reconstruction algorithms. Event collection procedures will also need to take account of the initial operation of the LHC machine which is expected to provide short concentrated bursts of data-taking separated by longer periods devoted to machine studies. In the beginning the total dataset will be small such that copies of the entire dataset of reconstructed events can be placed at Tier-1 or Tier-2 centers to allow for easy access by all members of the collaboration.

The LHC computing grid – an evolving service

The LHC computing grid is organized as a collaboration of the participating sites and the experiments, with agreements and decisions made by consensus. A few months before the start-up of the LHC accelerator the grid services were in operation with 140 active sites, handling over 300,000 new jobs per day, each taking up to a day or more to complete, and demonstrating the data transfer performance needed for the first year of operation. The workload was already widely distributed, with only 15% of the processing taking place at CERN and more than 50% at the Tier-2 sites. There are some very large sites and a large number of small sites, which fulfils the goal of enabling all sites, large and small, wherever they may be located, to participate effectively in the LHC analysis.

The computing service will continue to evolve throughout the life of the accelerator, growing in capacity to meet the rising demands as more data is collected each year. With the grid architecture that has been established funding agencies have considerable flexibility in deciding how and where to provide future computing resources, and the community is well placed to take advantage of new technologies – hardware, software and services – that may appear and that offer improved usability, cost effectiveness, or energy efficiency.

Further reading

Software Architecture and Tools

[1] L. Bass, P. Clements, and R. Kazman, *Software Architecture in Practice*, (2003) Addison-Wesley, ISBN: 978-0321154958.

[2] J. Lakos, *Large-Scale C++ Software Design*, (1996) Addison-Wesley, ISBN: 978-0201633627.

[3] G. Barand et al., "GAUDI - A software architecture and framework for building HEP data processing applications" *Computer Physics Communications* **140** (1) 45-55.

Grid computing

[4] I. Foster and C. Kesselman *The Grid: Blueprint for a New Computing Infrastructure* (1999) Morgan Kaufmann, ISBN: 1-558660-475-8

[5] Wikipedia: http://en.wikipedia.org/wiki/Grid_computing

[6] E. Laure, et al., *Programming the Grid with gLite*, EGEE Technical Report EGEE-TR-2006-001 http://cdsweb.cern.ch/search.py?p=EGEE-TR-2006-001

The LHC Computing Grid

[7] J. Knobloch (ed.), *The LHC Computing Grid Technical Design Report* CERN document CERN-LHCC-2005-024: http://lcg.web.cern.ch/LCG/tdr/LCG_TDR_v1_04.pdf

[8] M. Lamanna, "High-Energy Physics Applications on the Grid" in *Grid Computing: Infrastructure Service and Applications*, L. Wang et al., editors (2009), ISBN: 1420067664.

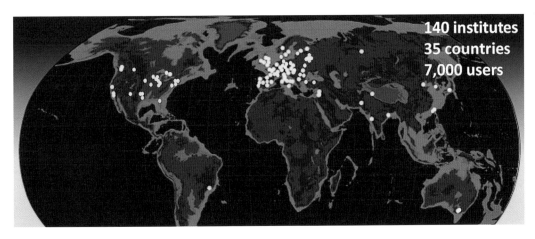

**140 institutes
35 countries
7,000 users**

72. The sites participating in the LHC Computing Grid.

Epilogue

On the morning of 10th September 2008, almost every office at CERN was deserted. The staff was packed into the many lecture rooms around the site to which images of the LHC control room were being transmitted. Little known to the people working in the control room, the same images were being beamed to TV stations all around the globe. The atmosphere was electric.

At precisely 9:30 in the morning, almost 15 years after approval, a beam was injected into the LHC. A bright spot was registered on a thin fluorescent screen at the entrance to the machine as the beam passed through it. An absorber block was move into the machine aperture at the end of the first octant and

The LHC control center on the morning of 10th September 2008.

the beam was observed to strike a second screen just upstream of the block on the very first shot, a journey of a little over 3 km. Over the next hour, the beam was coaxed around the ring from octant to octant. Finally, after the last absorber was removed, the beam made its first full revolution, evidenced by two small spots on the first screen, that of the injected beam and after it had made the first full 27 km revolution. A collective sigh of relief around CERN was almost audible. There were no obstructions in the vacuum chamber, as had been the case for a previous machine, and the polarities of the magnets guiding and focusing the beam seemed to be correct. This first step was followed not only by the CERN staff responsible for building the machine but also by an audience of millions of people around the world.

Two small spots signaled the completion of the first turn of a proton beam around the LHC.

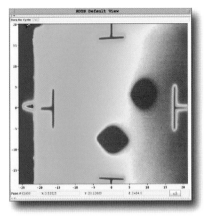

In the days that followed, rapid progress was made in getting a beam to circulate with very good lifetime. The Radio Frequency system was tuned to "capture" the beam, keeping it tightly bunched and it was centered in the vacuum chamber with remarkable precision. All bode well for a fast and smooth commissioning.

Then disaster struck. Before the 10th September injection test, seven of the eight octants had been tested up to the full energy of 5 TeV to be used for the first year of operation. For lack of time, the eighth octant had only been tested up to 4 TeV. As this last octant was being ramped up to its nominal energy, a flurry of alarms reached the control consoles and the safety systems were activated to protect the machine. Over the next few hours, it was found that the root cause of the problem was a failure of one of 50,000 soldered joints, which would have been a minor incident for a conventional machine but not for the LHC. Just warming up the faulty sector for repair would take 5-6 weeks, but it was found that there had also been quite considerable collateral damage due to over pressure in the helium circuits.

Once more, it was necessary to call on the resourcefulness and considerable resilience of the CERN staff. The first priority was to understand the sequence of events and to develop methods sensitive enough to detect any small anomaly elsewhere in the machine to avoid the possibility of such an event ever happening again. The second was to initiate the repair.

Now, nine month after these events, the LHC is ready to embark on its first year of operation. The construction of the machine and its detectors has been a monumental task involving literally thousands of people from all around the globe. It has pushed technology to the limit and has fostered International Collaboration on a grand scale, independent of race or religion in the interest of the quest for knowledge. As the LHC sails into unchartered waters, let us wish her "bon vent".

Lyndon Evans,
May, 2009
Geneva, Switzerland

Glossary

Absolute zero
Absolute zero of temperature corresponds to 0 K on the Kelvin scale or −273.15 °C on the Centigrade scale. Absolute zero can only be reached asymptotically. The lowest temperature ever reached is a few micro-Kelvin. The LHC operates at 1.9 K.

Accelerating cavity
Accelerating cavities produce the electric field that accelerates the particles inside particle accelerators. Because the electric field oscillates at radio frequency, these cavities are also referred to as radio-frequency (RF) cavities.

Antimatter
Every kind of matter particle has a corresponding antiparticle. Charged antiparticles have the opposite electric charge of their matter counterparts. Although antiparticles are extremely rare in the Universe today, matter and antimatter are believed to have been created in equal amounts at the Big Bang.

Boson
The collective name given to the particles that carry forces between particles of matter.

Calorimeter
An instrument for measuring the amount of energy carried by a particle. In particular, the electromagnetic calorimeter measures the energy of electrons and photons, whereas the hadronic calorimeter determines the energy of hadrons, that is, particles such as protons, neutrons, pions and kaons.

Center-of-mass energy
The center-of-mass energy is that energy available for producing new particles. In fixed target mode the centre-of-mass energy only increases as the square root of the energy of the projectile. In collider mode, it increases proportionally to the energy of the colliding projectiles.

Cherenkov radiation
Light emitted by fast-moving charged particles traversing a dense transparent medium faster than the speed of light in that medium.

Collider
Special type of accelerator where counter-rotating beams are accelerated and interact at designated collision points. The collision energy is twice that of an individual beam, which allows higher energies to be reached than in fixed target accelerators.

CP violation
A subtle effect observed in the decays of certain particles that betrays Nature's preference for matter over antimatter.

Cryogenic distribution line (QRL)
The system used to transport liquid helium around the LHC at very low temperatures. This is necessary to maintain the superconducting state of the magnets that guide the particle beam.

Cryostat
The cryostat is the thermally insulating device that houses the cold parts of the accelerator.

Dark matter / Dark energy
Only 4% of the matter in the Universe is visible. The rest is known as dark matter (26%), and dark energy (70%). Finding out what these consist of is a major challenge for modern science (Chap. 2).

Duoplasmatron
The source of all protons at CERN. The duoplasmatron ionizes hydrogen gas from which the protons are extracted by an electric field.

Electronvolt (eV)

A unit of energy or mass used in particle physics. One eV is extremely small, and units of a million electronvolts, MeV, or a thousand million electronvolts, GeV, are more common. The latest generation of particle accelerators reaches up to several millions of million electronvolts, the tera-electron volt (TeV). One TeV is about the energy of motion of a flying mosquito.

Forces

There are four fundamental forces in nature. Gravity is the most familiar to us, but it is the weakest. Electromagnetism is the force responsible for thunderstorms and carrying electricity into our homes. The two other forces, weak and strong, are confined to the atomic nucleus. The strong force binds the nucleus together, whereas the weak force causes some nuclei to break up. The weak force is important in the energy-generating processes of stars, including the Sun. Physicists would like to find a theory that can explain all these forces. A big step forward was made in the 1960s when the electroweak theory uniting the electromagnetic and weak forces was proposed. This was later confirmed in a Nobel-prize-winning experiment at CERN.

Gluon

The gluon is a special particle, called a boson, that carries the strong force, one of the four fundamental forces, or interactions, between particles.

Hadron

The hadron is a particle made of quarks contained by the strong force (Fig. 10 of Chap. 2). The two sub-families of hadrons are the baryons and the mesons. The proton is a member of the hadron family, thus the name Large Hadron Collider.

Hermetic (detector)

A particle detector sensitive to particles emitted at all angles from the interaction point (apart from the beamline direction).

Hadronic calorimeter

A layer of the standard particle detector, located before the muon chambers; it is used to absorb and measure the energy of particles mainly undergoing a hadronic interaction, such as pions, kaons, protons, and neutrons (Fig. 1 of Chap. 5).

Heavy-flavor physics

The study of properties of quarks with large mass and their decay products is called *heavy-flavor physics*, with an emphasis on the physics of the b and c quarks.

Higgs boson

The "Holy Grail of particle physics", the Higgs boson is the particle needed to complete the unified electroweak model. It is the particle corresponding to a field that permeates the universe, and, via its interaction with fermions and bosons, it allows them to acquire a mass, breaking the electroweak symmetry (Chap. 2). The ATLAS and CMS experiments have been designed with the goal of detecting the Higgs boson, if it exists.

High p_T (transverse momentum) measurements

In a proton collider, the collisions come from interactions of proton constituents, i.e., the quarks and gluons, each carrying only a fraction of the total proton energy. Most of the interesting events arise when the energy of the constituents is high; experimentally this results in the emission of particles with large momentum *perpendicular* to the colliding beam (p_T). High momentum in the direction *parallel* to the beam is not necessarily an indication of a high-energy collision.

Injector

The injector refers to the system that supplies particles to an accelerator. The injector complex for the LHC consists of several accelerators acting in succession (Chap. 1).

Interaction length

The interaction length refers to the average distance a hadron will travel before interacting with a given material. A large number of interaction lengths means that hadrons will be effectively stopped or filtered out by that material.

ISR

The Intersecting Storage Rings, which was the first proton-proton collider built at CERN. It operated from 1971 to 1984.

Kaon

A meson containing a strange quark (or antiquark). Neutral kaons come in two kinds, long-lived and short-lived. The long-lived ones occasionally decay into two pions, a CP-violating process. (See also Particles.)

Klystron
A klystron is a linear-beam vacuum tube that functions as a high-gain radio-frequency amplifier. The klystrons supply the radio frequency that accelerates the LHC proton beam.

Lambda point
The temperature (2.17 K) at which liquid helium makes the transition to the superfluid state.

LCG (LHC Computing Grid)
The mission of the LCG is to build and maintain a data-storage and analysis infrastructure for the entire high-energy physics community that will use the LHC.

LEP
The Large Electron–Positron Collider, which ran at CERN until 2000.

Luminosity
The luminosity refers to the average beam density at the collision point. It is expressed as the number of particles per unit area per unit of time, and thus has the units $cm^{-2}\ s^{-1}$. It is the parameter of an accelerator that indicates the production rate of the processes to be observed.

Magnetic fields (toroidal and solenoidal)
A charged particle moving in a constant magnetic field will undergo a screw-shaped trajectory, with its radius of curvature proportional to its momentum. Modern particle detectors are immersed in strong magnetic fields, created by large electro-magnets in the *solenoid* or *toroid* configuration. A solenoid has a cylindrical shape; the magnetic-field generating current flows along its side, perpendicular to its axis, producing uniform magnetic field lines

that run parallel to the axis. (Fig. 14 of Chap. 5) The other approach is to use toroid magnets (more "doughnut"shaped, see Fig. 28 of Chap. 5) through which current is flown to create magnetic field lines perpendicular to the beam axis.

Minimum-bias processes
Most of the interactions in a hadron collider occur between low-energy quarks and gluons of the proton and are lacking the energy to produce high-mass final states (the outcome of interest). The trigger and data acquisition system of the detectors is optimized to filter out the data related to these minimum-bias processes by selecting final states with large transverse momentum.

Missing transverse energy (or momentum)
In particle collisions, momentum is conserved. In proton-proton interactions, the hard collision is generated by quarks and gluons, whose energy is unknown for the single event, and part of the energy is emitted (and not measured) in the direction of the beam. Thus, as seen by the detectors, momentum conservation is only meaningful in the transverse direction (perpendicular to the beam axis). Any indication of missing transverse momentum indicates that undetectable particles, such like neutrinos, have been produced.

Multiple scattering
A charged particle passing though material will be influenced by the electric field of the material's atoms and undergo random deviations from its original trajectory, even in the absence of hard collisions with the nuclei of the material. This *multiple scattering* phenomenon decreases the precision of the reconstruction of the particle's direction and initial momentum.

Occupancy
The occupancy of a detector is the average fraction of readout channels that produce a non-zero signal following an event.

Particles
There are two groups of elementary particles, quarks and leptons (Chap. 2). The quarks are up and down, charm and strange, top and bottom. The leptons are electron and electron neutrino, muon and muon neutrino, tau and tau neutrino. There are four fundamental forces, or interactions, between particles, which are carried by special particles called bosons. Electromagnetism is carried by the photon, the weak force by the charged W and neutral Z bosons, the strong force by the gluon; gravity is probably carried by the graviton, which has not yet been discovered. Hadrons are particles that feel the strong force. They include mesons, which are composite particles made up of a quark–antiquark pair and baryons, which are particles containing three quarks. Pions and kaons are types of meson. Neutrons and protons (the constituents of ordinary matter) are baryons; neutrons contain one up and two down quarks; protons two up and one down quark.

Parton
A parton is any "point-like" constituent of the hadrons; known partons are quarks, antiquarks and gluons.

Photon
The force carrier particle of electromagnetic interactions.

Pion
The pion is a meson composed of a quark-antiquark pair of the first family (up and down) and are the least massive type of mesons. Pions can be charged or neutral, and are the most abundant particles in hadronic jets.

Positron
The positron is the antimatter partner of the electron. It therefore shares many of its properties (e.g., mass), but it is of opposite electric charge.

PPBARS
The CERN Proton-Antiproton collider that discovered the W and Z bosons in 1983.

PS
The Proton Synchrotron, backbone of CERN's accelerator complex.

PSB
The PS Booster synchrotron is the intermediate stage of acceleration between the Linac and the PS.

Pseudorapidity
A spatial coordinate describing the deviation of a particle with respect to the axis of the beam pipe. It is related to the polar angle of the trajectory.

Quadrupole magnet
A magnet with four poles, used to focus particle beams rather as glass lenses focus light. There are 392 main quadrupoles in the LHC.

Quantum electrodynamics (QED)
The theory of the electromagnetic interaction.

Quantum chromodynamics (QCD)
The theory for the strong interaction, analogous to QED.

Quark–gluon plasma (QGP)
A new kind of plasma in which protons and neutrons are believed to break up into their constituent parts. A QGP is believed to have existed just after the Big Bang.

Quench
A quench occurs when a small part of the magnet superconductor is heated above its transition temperature. Having acquired resistivity, additional heating will take place, leading to an event during which a large part of the superconducting magnet may quickly and uncontrollably warm. Damage is avoided with equipment and procedures that safely dissipate the energy stored in the magnet when a quench occurs.

Radiation length
For particles that interact mainly through electromagnetic forces, the radiation length is the average distance the particle will travel in a given material. This parameter is applied to photons and electrons.

Ring Imaging Cherenkov (RICH) counter
A kind of particle detector that uses the Cherenkov light emitted by fast-moving particles as a means of identifying them.

Roman-pot technique
The Roman-pot technique is used by certain detectors that measure the scattering of the beam protons at angles very close to the beam axis; to do this, the detection scream is retracted during beam injection, acceleration and dumping (when the beam is broadest), and closed into its operational position only once stable (and narrow) colliding beams are obtained (Fig. 5.54).

Scintillation
The flash of light emitted by an electron in an excited atom falling back to its ground state.

Sextupole
A magnet with six poles, used to apply corrections to particle beams. At the LHC, eight- and ten-pole magnets will also be used for this purpose.

SPS
The Super Proton Synchrotron. An accelerator that provides beams for experiments at CERN, as well as preparing beams for the LHC.

Strong force
The strong force holds quarks together within protons, neutrons and other particles. It also prevents the protons in the nucleus from flying apart under the influence of the repulsive electrical force between them (because they all have positive charge). Unlike the more familiar effects of gravity and electromagnetism where the forces become weaker with distance, the strong force becomes stronger with distance.

Superconductivity

A property of some materials, usually at very low temperatures, that allows them to carry electricity without resistance. If you start a current flowing in a superconductor, it will keep flowing forever – as long as you keep it cold enough.

Superfluid

Superfluidity is a state of some liquids, notably liquid helium, that have been cooled to the point that it has lost its viscosity (i.e. mechanical resistance). The LHC exploits the superfluid properties of helium for the cooling of the cryomagnets.

Supersymmetry

A theory that predicts the existence of heavy 'superpartners' to all known particles. It will be tested at the LHC.

Synchrotron radiation

The light emitted when electrons or positrons (and other charged particles) are bent in a magnetic field.

Tesla

The unit of magnetic field strength. A conventional electromagnet operates at a maximum field of about 1.8 tesla. The LHC dipoles operate at 8.3 tesla thanks to superconductivity and helium superfluidity.

Transfer line

Transfer lines carry beams of particles, e.g., protons, from one accelerator to another using magnets to guide the beam.

Transition radiation

When a charged particle travelling close to the speed of light crosses the boundary between two media with different indices of refraction, electromagnetic radiation is emitted (the transition radiation). At LHC energies, the presence of this radiation helps discriminate between very light (and fast) electrons from other charged particles, such as pions.

Trigger

Trigger refers to an electronic system for spotting potentially interesting collisions in a particle detector, which then signals to the detector's read-out system that the data resulting from the collision should be recorded.

Vertex and secondary vertex

The interaction point in space of each individual collision is called the *primary vertex*. Some of the particles created in the collision will have a very short lifetime, such as b mesons or tau leptons, and these will in turn decay after having flown a few millimeters away from the vertex. A high-precision tracker can distinguish the new particles produced at this *secondary vertex* from those produced at the primary vertex.

W and Z bosons

The W and Z bosons are the elementary particles that mediate the weak force. They were first discovered in the UA1 and UA2 detectors at the PPBAR collider in 1983. The W weighs 80.4 GeV and the Z weighs 91.2 GeV

Weak force

The weak force acts on all matter particles and leads to, among other phenomena, the decay of neutrons (which underlies many natural occurrences of radioactivity) and allows the conversion of a proton into a neutron (responsible for hydrogen burning in the centre of stars).

About this glossary:
Many of the terms have been adapted from CERN sources (Copyright CERN 2008 - Web Communications, DSU-CO), with new or enriched entries supplied by Dr. Mario Campanelli.